INTERDEPENDENT AND UNEVEN DEVELOPMENT: GLOBAL-LOCAL PERSPECTIVES

Also in the series:

The Industrial Enterprise and Its Environment: Spatial Perspectives
Edited by Sergio Conti, Edward J. Malecki and Päivi Oinas
ISBN 1 85628 876 5

The Asia Pacific Rim and Globalization: Enterprise Governance and Territory
Edited by Richard Le Heron and Sam Ock Park
ISBN 1 85628 894 3

Environmental Change: Industry, Power and Policy
Edited by Michael Taylor
ISBN 1 85972 161 3

Interdependent and Uneven Development

Development

Global-local perspectives

Edited by
MICHAEL TAYLOR
University of Portsmouth, UK

SERGIO CONTI
University of Turin, Italy

Aldershot · Brookfield USA · Singapore · Sydney

Published by
Ashgate Publishing Ltd
Gower House
Croft Road
Aldershot
Hants GU11 3HR
England

Ashgate Publishing Company
Old Post Road
Brookfield
Vermont 05036
USA

HF
1021
.I57
1997

British Library Cataloguing in Publication Data

Interdependent and uneven development : global-local
        perspectives. - (The organisation of industrial space)
        1. Economic geography 2. Dependency 3. International economic
        integration
        I. Taylor, Michael II. Conti, Sergio
        330.9

Library of Congress Catalog Card Number: 97-72314

ISBN 1 85972 445 0

Printed in Great Britain by The Ipswich Book Company, Suffolk.

Contents

PART II Business enterprise, local networks and 'spatial learning'

PART III Labour, locality and restructuring

List of figures

List of tables

List of contributors

Bjørn T. Asheim, Department of Sociology and Human Geography, University of Oslo, Norway.

Sergio Conti, Dipartimento Interateneo Territorio, University of Turin, Italy.

Ian Cooper, Department of Geography, University of Auckland, New Zealand.

Federico Cuñat, Departement de Sciences Humaines, Cité Scientifique, Ecole Centrale de Lille, France.

Carol Ekinsmyth, Department of Geography, University of Portsmouth, United Kingdom.

Robert Fagan, Department of Geography, School of Earth Sciences, Macquarie University, New South Wales, Australia.

Wolf Gaebe, Geographisches Institut, The University of Stuttgart, Germany.

David Hayward, Department of Geography, University of Auckland, New Zealand.

Andrew E.G. Jonas, Department of Geography University of Hull, United Kingdom.

Richard Le Heron, Department of Geography, University of Auckland, New Zealand.

Simon Leonard, Department of Geography, University of Portsmouth, United Kingdom.

Edward J. Malecki, Department of Geography, University of Florida, USA.

Anders Malmberg, Department of Social and Economic Geography, Uppsala University, Sweden.

Phillip M. O'Neill, Department of Geography, University of Newcastle, New South Wales, Australia.

Martin Perry, Department of Geography, National University of Singapore, Singapore.

Örjan Sölvell, Institute of International Business, Stockholm School of Economics, Sweden.

Michael Taylor, Department of Geography, University of Portsmouth, United Kingdom.

Bernadette Thomas, Departement de Sciences Humaines, Cité Scientifique, Ecole Centrale de Lille, France.

Deborah M. Tootle, Department of Geography, University of Florida, USA.

1 Introduction: perspectives on global-local interdependencies

Michael Taylor and Sergio Conti

Introduction

This book is primarily concerned with the geographical implications of global economic change and the uneven impact that processes of globalisation and economic interdependence have in local communities. There is no doubt that the structure of capitalism has been drastically reshaped in the late twentieth century. The expansion of transnational corporations and transnational capital have been unprecedented in the period since 1945. Flows and transfers of materials, money and goods are increasingly within and between transnational corporations or controlled by them. Economic restructuring and liberalisation have, through austerity measures, pushed Third World countries close to the economic brink, torn and dismembered the economic and social fabric of most of the socialist states of the Second World, and 'red-lined', abandoned and excluded entire places, communities and social groups in the so-called First World economies of the OECD. These changes challenge the foundations of explanation in economic geography as we try to understand the dynamic spatial patterning of production and consumption. At best we are now groping towards new forms of explanation and understanding. At worst we are still struggling to cast off the straight jacket of obsolete approaches.

This book can hope to do no more than throw light on a small range of issues as we forage in the dark for new paths towards understanding the shifting patterns of globalisation. It draws together eleven contributions on the evolving global-local dialectic from an essentially First World perspective in order to reflect on global-local interdependencies and connectivity, industrial districts, knowledge and 'local learning', and the nature and operation of localities and local labour makets in late twentieth century capitalism.

New global geographies

Economic change resulting from the interplay of, on the one hand, crisis and, on the other hand, stability (Ekinsmyth et al, 1995) is now creating new geographies of economic activity and six such geographies can be identified associated with economic globalisation.

In production, the global reach of the transnational corporation (Barnet and Müller, 1975) has, through ownership, strategic alliance and the power of subcontracting, created a 'Nike world economy' of globally integrated integrated and coordinated production which goes much further than simply internationalised production, and is also said to be qualitatively very different (Dicken et al, 1995; Nilsson et al, 1996; also see Conti and Asheim in this volume).

Financial and business service organisations initially stalked the industrial TNCs on to the global scene, aided by the commodification of money and financial deregulation from the 1970s on, although they are now as much the stalked as the stalkers in the global economy (Corbridge, Thrift and Martin, 1995). They have created a global financial system hubbed on a few, key global cities tied into an electronic web.

Intimately associated with these new global patterns are equally important emerging patterns of global control. Here there are the shifting patterns and processes of control within large corporations (Handy, 1994), the emergence of new trading blocs (such as the EU, NAFTA, ASEAN and APEC, for example), and the creation of the 'competition state' (Dicken, 1994).

Intertwined with these changing geographies of production and control are the geographic possibilities moulded by the enabling technologies of IT, telecommunications and satellites coupled with the transportation technologies of aviation and bulk transport (Dicken, 1992). These are the geographies of time-space compression and substitutability (material for material, source for source, or labour for labour) which are as important for their characteristics of exclusion as much as they are for their potential for global inclusion.

New global geographies are also emerging in capital-labour relations, and these find their expression in the new international division of labour, global patterns of labour migration, the re-establishment of the 'local' in understanding the functioning of labour markets (Peck, 1996), and the discourses of control built on the disempowering 'global' (see O'Neill in this volume).

Finally, there are the emerging global geographies of consumption - the 'McDonaldisation' of society. Here there is the hegemony of transnational culture and the vexed issues of inclusion and exclusion not only at the scale of the North versus the South but also between one section of society and another and between one place and another.

These six sets of patterns are emerging geographies of point and counterpoint between the global and the local. They are geographies of consumption versus production; the use of funds versus the source of funds; control versus resistance and autonomy; and inclusion versus exclusion. All are indicative of global-local tensions

- and, indeed, a global-local dialectic - and all are targets of enquiry in an economic geography of globalisation.

The global-local dialectic

Globalisation is usually interpreted as involving a global-local dialectic '... where local events constitute global structures which then impinge on local events in an iterative continuum' (Taylor et al, 1995a, p. 9, drawing on Watts, 1992). The big question however is not whether this dialectic exists but how intense the processes are that constitute these interrelationships. This is hotly contested terrain (see Amin, 1997). At one extreme is the view that globalisation is the triumph of unstoppable global capital over local autonomy and identity. By this view, transnational corporations have slipped the shackles of the nation state and, in turn, the nation state has been hollowed out. Time-space compression has turned the world economy into a dense, massively complicated and indecipherable web of interconnections and accelerating flows of materials, goods, finance, information and money (Harvey, 1989; Castells, 1993). Places, people and communities are victims on the rack of international capital: they are powerless - as are their governments. An intermediate position is that globalisation is much more benign and involves no more than a blurring of traditional territorial and social values. By this view time-space compression is not new and was, in fact, decried just as intensively in the railway era as it is now in the electronic era. Trade and finance too are said to have been as international before the First World War as they are now, and it was the inter-war years that were the aberrant time of insularity and localism. The state now must find a new role and mode of operation for, 'without the state, capitalism's creative destruction ... degenerates into private mafias ... and short-term economic destruction' (Taylor et al, 1995b, p. 383). At the other extreme of the interpretative spectrum, the extent and effect of globalisation are seen as a massive exaggeration. By this view globalisation is no more than intensified international exchange between quite separate national and social entities. Regionalisation and the formation of trading blocs like the EU, NAFTA, ASEAN and the latent APEC grouping are seen as far more significant than the homogenising forces of globalisation.

This contested terrain of globalisation is however no more than a debate over variously interpreted trends and tendencies and, as Amin and Thrift (1995) have pointed out, these trends and tendencies in no way constitute a finished theory of globalisation. The debate revolves around three key issues:

* what is the nature and shape of the relationships that bind actors, agents, places and communities into international and global systems of interconnection? Are they large and complex networks that are all-pervasive and robust, or are they small, personalised and embedded networks that are fragile and ephemeral? Indeed, are they networks?

3

- where does power lie a globalised systems of global-local interrelationships? Asymmetry is said to be an essential element of network forms (Dicken and Thrift, 1992; Taylor, 1995), but what of trust and reciprocity, 'power to' and 'power over', and the question of resistance?
- how do 'space' and 'place' fit into these systems of asymmetries and interconnectedness? Is geography a game board on which global games are played? Are 'places' and localities victims to be used and abused by a placeless capitalism? Or are 'space' and 'place' intrinsic to the actions of the agents embroiled in the game of dynamic capitalism?

Central to these issues is the nature and exercise of power between agents within a capitalist system. As it was put by Dicken (1994) 'the "global-local nexus" ... is far more than merely a question of the geographic scale at which economic processes occur. More fundamentally, it is a question *of where power lies*, and it is a central problematic facing both firms and states' (p. 102, emphasis added). Unfortunately, the notion of power relations expressed in this way is very much a caricature of the powerful, empowered 'global' versus the powerless victim 'local' - a view that has become an all too frequently intoned mantra. In what appears to be a search for a grand theory of globalisation, what is currently missed in many analyses of the global-local nexus is the detailed intricacy of the nature and exercise of power in the political-economic arena of present day capitalism.

The search for theory: 'regulation' versus 'systems'

At the heart of this global-local debate in economic geography and geographic perspectives on the exercise of power are two very different conceptualisations; on the one hand, a broadly *regulation approach* and, on the other hand, a broadly *systems approach.*

Regulation theory is built on French-inspired historiography in which the history of capitalism is understood as a series of periods within which specific institutional forms are privileged which determine the organisational characteristics of a particular mode of production and its associated social relations. As an original synthesis of a number of theories embracing Keynesian economics, institutionalist ideas and Marxist analysis it tends to betray its origins. Its emphasis is on social relationships derived from orthodox Marxism, though it rejects the idea of an unvarying long-run dynamic from the same root. Based on capital-labour relationships, social relations of production are framed within market mechanisms. Since these mechanisms are specific to each society and historical phase, time-specific and space-specific structural and institutional forms arise to regulate the process of accumulation and the manner of individual and collective agency. Thus, the realisation of capitalist accumulation can not be divorced from the specific situations within which it is 'embedded', countering the technological determinism of Schumpeterian views of change.

The systems approach, in its most modern version, goes beyond the traditional organic interpretation of open systems by developing a theoretical schema within which openness and closure are inseparable components of an interpretative whole. Simply put, a system is open to the energy it needs to keep it alive but closed in the form of its organisation in order to protect it from the disturbing and disorganising forces of its environment. Systems in these terms are self-referential and self-organising. Thus, the organisation of a system and the structural transformation it encounters depends on circular relationships which express themselves through networks of interconnection and give it identity. The system is no longer an organic focus of attention, as in the structural contingency model in organisation theory for example. It can now be disassembled into subsystems (systems in themselves) in each of which the borders are determined by processes of organisational closure that provide each subsystem with a unique identity. When it is transposed into the spatial arena, the intermediate entity of the partial system (subsystem) becomes a regional or local economy rather than the system as a whole which in macro terms may be national, world-wide or global. The theory of autopoesis introduces into this framework the possibility of distinguishing and characterising a system in terms of its organisational and regional identity making the observer's point of reference *internal* to the system rather than *external* as was the perspective in earlier versions of open systems theory.

Certainly, both of these theoretical concepts are substantially different to those derived from neo-classicism, neo-positivism, functionalism and Marxist structuralism. These latter approaches to theory largely adopted a mechanistic world view which increasingly appears less relevant in a world of intensifying competitive and technological complexity. As categories of 'simplicity' lose their significance in theory building, system dynamics become interpretable as complex relationships between the global and the local, i.e. they present a social-economic world that can be understood in terms of relationships and can be represented as networks. Notwithstanding this element of congruence, the regulation and systems perspectives on global-local interconnection are radically different views on economic processes. In the regulation approach, general global processes are considered as dominant over local phenomena, and the 'local' assumes the position of a regulatory entity only at a level (which is usually that of the nation state) where the institutional conditions for the creation of competitive advantage are present. The epistemological foundation of the systems perspective is, however, profoundly different. The 'local' under these circumstances becomes a protagonist in its own right. It is itself a complex totality capable of autonomous behaviour endowing it with its own identity which, in turn, serves to distinguish it from, its environment and other systems.

Complexity and 'weak explanation'

However, it has been the intersection of these very different perspectives - one from the 'global' and one from the 'local' - that have highlighted important facets of the global-local nexus. Two such facets stand out; the idea of the *territorial embeddedness* of enterprise and idea of *complexity* undermining 'hard' categories of knowledge and favouring 'weak explanation'. Recognising embeddedness involves a rediscovery of territoriality and the analytical category 'local' (understood as a complex, environmental totality) (see Grabher, 1993). Recognising complexity casts doubt on the long held conviction that the range and variety of economic and human activities can be explained in simple, universal terms, which make it possible to forecast the historical succession of events through the rules of quantification and economism.

The rethinking of the explanatory categories of traditional approaches to economic explanation has occurred across the social sciences as a whole as well as in geography, and the rediscovery of territoriality and 'weak explanation' are two important outcomes of recent debate. In the economic arena in particular, attention has been refocussed on the social as well as the economic characteristics of territorial systems. A major outcome has been the reassertion of the role of the original Marshallian concept of 'industrial atmosphere' in economic change which serves to leaven purely economic phenomena with community dynamics established by long-term, historical-cultural processes. Indeed, the debate over the significance of non-economic components in understanding the dynamics of districts' economic environments has had important theoretical consequences. It has undermined some of the basic certainties of conventional economic theory, eating away at their methodological and ideological purity. It has allowed the territorial system to be recognised as an intermediate entity (located between the actor and the complex system) dialectically connected to global processes.

The economic system can no longer be viewed as a single organic whole regulated by universal laws despite its economic and social diversity. Reality is now recognised as being far more complex. It comprises many diverse, autonomous and interrelated systems which are, at one and the same time, a celebration of the 'local' and a recognition of the 'non-reducibility' and 'irrepeatability' of observable phenomena (see Dupuy, 1985).

The 'global', therefore, is to be understood in a relational sense. Its extent is not definable a priori but only in terms of the relationships that interconnect its constituent subsystems. The global, in other words, is constituted of characteristics of the systems it connects, modelled upon their specific configurations. The 'local' too is not a simple physical entity. It is not simply a part of a complex system, but is a whole in itself, endowed with its own identity which distinguishes it from the environment and from other systems. It is composed of agents who are aware of this identity and who are capable of autonomous collective behaviour. This is, therefore, a system which interacts with the outside according to its own rules which are largely informal but which are still sufficient to guarantee its reproduction over time.

The global-local nexus elaborated in this way is unequivocally complex. At its heart is the concept of the *network* which has gained considerable prominence in recent economic geographic literature (see Dicken and Thrift (1992), Grabher (1993), Dicken (1994), Conti and Dematteis (1995), Camagni (1991), Perrin (1990) and Rallet and Torre (1995), for example). The network here is interpreted as a representation of social interaction between actors - a metaphoric meaning quite different to that of conventional neo-positivist theories. A system seen in this light evolves and expresses itself through a relational dynamic involving agents who act both collectively and individually. Thus, local actors or agents interact globally not only as individuals but also as an expression of the territorialised socio-economic relationships within which they are embedded. At the same time, the same local systems interact with other systems at other levels through the intermediation of the individual agents that belong simultaneously to both local and global networks.

What we are left with, then, is a perspective on global-local interconnections as a mass of complexity and contingency fuelled by recursive relationships - indeed relativism writ large.

This need to recognise complexity together with the previously identified need to come to grips with the 'place' and 'circumstance' of the exercise of power has been addressed with great clarity by Thrift (1995) in his analysis of economic globalisation as hyperactivity. He begins his analysis with the development of a powerful critique of the illegibility of the labyrinthine complexity of globalisation, the hypermobility of globalisation as a space of flows and the drama of time-space compression to suggest that we need to change the way we do theory. Persuasively he argues that we need to be more tentative and less exaggerated in our views, especially on globalisation. The large economic changes that dominate current portrayals of globalisation are seen as complex and contingent and more appropriately interpreted as tendencies that emerge from a multiplicity of minor processes that repeat, imitate and reinforce one another, leading only gradually towards convergence (p. 32-34). The functional relationships binding agents are by this interpretation incomplete, tentative and approximate, 'constantly ... ordering a somewhat intractable geography of different and often very diverse geographical contexts' (Thrift, 1995, p. 33). New forms of connection are made and destroyed just to create networks 'that will last a little longer' (p. 33). Thrift goes on to build on the work of Latour (1993) interpreting networks as essentially small and social rather than global and economic, re-crafting 'global' networks as *skeins* of networks which are 'neither global nor local but are more or less long and more or less connected' (Latour, 1993, p. 122). He maintains that to think otherwise is to mistake length and connection for differences in scale level. The contingency and imprecision that is fundamental to this way of thinking is labelled 'provisonality' by Thrift (1995, p. 35). There are by this view no big answers in the messy contingent world, only fleeting glimpses of possible answers.

The structure of the volume

Using 'provisonality' as a jumping off point, what is provided in this volume is, in essence, a set of fleeting glimpses of global-local interrelationships from a range of theoretical perspectives: from the neo-positivist to the structuralist and the regulationist to the post-modernist.

Part I comprises four chapters that explore issues of global-local connectivity. Conti explores and begins to unpack the global-local dialectic in a review of concepts and theoretical proposals. The central plank of the argument developed in this chapter is that the industrial district when viewed as a system is more than a mere representation the 'local' but is, by being a socio-cultural microcosm of networked relationships, a window on the dialectic between the 'global' and the 'local' under conditions of increasing globalisation and complexity in the contemporary world economy. Taylor, Ekinsmyth and Leonard argue that to more fully appreciate global-local interdependencies and tensions, notions of 'place', 'space' and 'territory' need to be unpacked to recognise the nature of the spatialities of the actors and agents involved. Their concern is with the spatiality of interconnection which is interpreted as neither a spatial game board across which aspatial actions are played nor a spatial process in its own right. The spatiality of interconnection is, rather, *spatiality within actions* - i.e. within the actions of individual actors and collective agencies - which almost invariably involves tension and conflict and which may produce temporarily stable geographies (spatial fixity) only through largely socially generated stabilising processes. Le Heron, Cooper, Hayward and Perry explore global-local connectivity through an examination of the global eco-commodity system as it impacts in the horticultural sectors of the Hawkes Bay region in New Zealand. They develop an integrated conceptualisation of commodity chains and explore in detail the pressures that they exhert on the export orientated horticulture of one particular locality. In the last chapter in this section, Gaebe further elaborates global-local connectivity, but at the level of the large business enterprise and the TNC, through an examination of the use of strategic alliances between firms to secure and regain competitive advantage. The point and counterpoint of corporate strategy is reviewed to elaborate the functional and spatial outcomes of alliance building.

The four chapters of Part II approach the issue of global-local interdependence from the perspective of local networks and 'spatial learning'. Malmberg and Sölvell focus on the role of agglomeration in generating a firm's ability to create and sustain competitiveness under conditions of economic globalisation. They explore the issues of agglomeration through a close examination of local knowledge accumulation, the localised nature of innovation, barriers to the diffusion of knowledge, the attraction of resources to spatial economic clusters, and the linking of TNCs into localised processes of knowledge creation. Asheim takes this issue much further and interprets the continued relevance and success of industrial districts in an era of economic globalisation as being dependent upon the collective learning capacities of the populations of SMEs within them and the innovativeness and flexibility that derives from the learning-based competitiveness of 'learning regions'. Cuñat and Thomas extend the analysis of local networks and embeddedness to embrace the distinctive local socialisation processes that have occurred in the textile sector of northern France. Here the pressures and problems of restructuring coupled with the establishment of subcontract production have seen the recent emergence of a new socially based collective dynamic between the large corporate distributors, employer organisations and the subcontract manufacturers. The dynamic is based on a formal charter incorporating all the parties, which socialises relationships far beyond what would be expected in pure market exchange. Finally in Part II, Malecki and Tootle examine relatively newly established networks of SMEs in the US to reflect on issues of embeddedness, local economic cultures and local growth.

Part III comprises three chapters in which global-local interdependencies are approached from the perspective of the locality and the local labour market. Fagan uses people's experiences in the outer suburban labour market of Western Sydney to understand contemporary suburban employment experiences in a globalising world. He suggests that the ideology of globalisation is as important as a source of change owing to its role in shaping the economic and social policies of the state as are the measurable impacts of global change itself. Labour control regimes are the focus of the chapter by Jonas. He argues persuasively that neither a global nor a local perspective provides an appropriate perspective from which to view contemporary developments in the geography of labour control. Instead he argues that each highlights elements of intrinsically uneven and changing geographies which he illustrates using US case studies. Finally, O'Neill examines the decline of the Newcastle steelworks in New South Wales, Australia, to investigate the meanings of internationalisation that are attached to the restructuring process. He weaves together the detail and complexity of global-local interdependencies in terms of crisis, rhetoric, restructuring, residual value and discourses of control.

References

Amin, A. (1997), 'Placing Globalisation', paper presented to the RGS-IBG Annual Conference, Exeter, 7-9 January.

Amin, A. and Thrift, N. (1994), 'Living in the global', in Amin, A. and Thrift, N. (eds) *Globalization, Institutions, and Regional Development in Europe*, Oxford University Press, Oxford, pp. 1-22.

Barnet, R. and Müller, R. (1975), *Global Reach. The Power of the Multinational Corporations*, Jonathan Cape, London.

Camagni, R. (1991), *Innovation Networks, Spatial Perspectives*, Belhaven-Pinter, London.

Castells, M. (1993), 'The information economy and the new international division of labour', in Carnoy, M. et al (eds), *The New Global Economy in the Information Age*, Pennsylvania State University Press, University Park, PA.

Conti, S. and Dematteis, G. (1995), 'Enterprises, systems and network dynamics', in Conti, S., Malecki, E.J. and Oinas, P. (eds), *The Industrial Enterprise and Its Environment: Spatial Perspectives*, Avebury, Aldershot, pp. 217-237.

Corbridge, S, Thrift, N. and Martin, R. (eds) (1994), *Money, Power and Space*, Basil Blackwell, Oxford.

Dicken, P. (1992), *Global Shift: The Internationalization of Economic Activity*, Paul Chapman Publishing, London.

Dicken, P. (1994), 'The Roepke Lecture in economic geography; global-local tensions: firms and states in the global economy', *Economic Geography*, vol. 70, pp. 101-128.

Dicken, P., Forsgren, M. and Malmberg, A. (1994), 'Local embeddedness and transnational corporations', in Amin, A. and Thrift, N. (eds) *Globalization, Institutions, and Regional Development in Europe*, Oxford University Press, Oxford, pp. 23-45.

Dicken, P. and Thrift, N. (1992), 'The organization of production and the production of organization: why business enterprises matter in the study of geographical industrialization', *Transactions of the Institute of British Geographers*, New Series, vol. 17, pp. 279-291.

Dupuy, G. (1985), *Ordres et désordres. Enquête sur un nouveau paradigme*, Deoul, Paris.

Ekinsmyth, C., Hallsworth, A., Leonard, S. and Taylor, M. (1995), 'Stability and instability: the uncertainty of economic geography', *Area*, vol. 27, pp. 279-289.

Grabher, G. (eds) (1993), *The Embedded Firm: On the Socioeconomics of Industrial Networks*, Routledge, London.

Handy, C. (1994), *The Empty Raincoat: Making Sense of the Future*, Hutchinson, London.

Harvey, D. (1989), *The Condition of Postmodernity*, Basil Blackwell, Oxford.

Latour, B. (1993), *We Have Never Been Modern*, Harvester Wheatsheaf, Brighton.

Nilsson, J.-E., Dicken, P. and Peck, J. (eds) (1996), *The Internationalization Process: European Firms in Global Competition*, Paul Chapman Publishing, London.

Peck, J. (1996), *Work-place: The Social Regulation of Labor Markets*, The Guilford Press, New York.

Perrin, J.-C. (1990), *L'environnment des enterprises innovantes: réseaux et districts*, Centre d'Economie Régionale, Publication no. 115, Aix-en-Provence.

Rallet, A. and Torre, A. (1995), *Economie industrielle et économie régionale*, Economica, Paris.

Taylor, M. (1995), 'The business enterprise, power and patterns of geographical industrialisation', in Conti, S., Malecki, E. and Oinas, P. (eds) *The Industrial Enterprise and Its Environment: Spatial Perspectives*, Avebury, Aldershot, pp. 99-122.

Taylor, P.J., Watts, M.J. and Johnston, R.J. (1995a), 'Global change at the end of the twentieth century', in Johnston, R.J., Taylor P.J. and Watts, M.J. (eds), *Geographies of Global Change: Remapping the World in the Late Twentieth Century*, Blackwell, Cambridge, MA, pp. 1-10.

Taylor, P.J., Watts, M.J. and Johnston, R.J. (1995b), 'Remapping the world: What sort of map? What sort of world?', in Johnston, R.J., Taylor P.J. and Watts, M.J. (eds), *Geographies of Global Change: Remapping the World in the Late Twentieth Century*, Blackwell, Cambridge, MA, pp. 377-385.

Thrift, N. (1995), 'A hyperactive world', in Johnston, R.J., Taylor P.J. and Watts, M.J. (eds), *Geographies of Global Change: Remapping the World in the Late Twentieth Century*, Blackwell, Cambridge, MA, pp. 18-35.

Watts, M.J. (1992), 'Capitalisms, crises and cultures, I: notes toward a totality of fragments', in Pred, A. and Watts, M.J. (eds), *Reworking Modernity*, Rutgers University Press, New Brunswick, NJ.

Part I: Global-local connectivity

Global-local interdependence is more than a question of the geographic scale at which economic processes occur, it is a question of connectivity and the linking of actors and actions in places and through time. The four chapters of this part of the volume approach this issue of connectivity from very different vantage points to provide fresh glimpses of the processes and trajectories of change. Conti boldly grasps the theoretical nettle to explore new systems theoretic and network approaches, embracing industrial districts and embeddedness, as a lens through which to observe the global-local dialectic. Taylor, Ekinsmyth and Leonard begin with the individual as their point of reference and explore the conflicting spatialities that individuals bring to the connectivities of global-local interdependence. Le Heron, Cooper, Hayward and Perry make their observations from yet another vantage point - that of the global eco-commodity system and local efforts to integrate into it using quality management techniques. Finally, Gaebe addresses the issue of connectivity in globalisation from the perspective of the large business organisation and the strategic alliances they enter into to gain, maintain, steal and protect competitive advantage on this world-wide stage.

2   Global-local perspectives: a review of concepts and theoretical proposals

Sergio Conti

Introduction

Globalisation is not a new phenomenon, if by the term one means the processes of *internationalisation* which over past decades have characterised the involvement of the major industrial corporations in the growing integration of national economies through flows of technical know-how, raw materials, intermediate goods and final products and services.

In that logic, the *multi nationalisation* of 'national' companies - through the creation of local branches, the acquisition of already existing firms, or through commercial, financial and technological agreements - overlapped with a conception of the enterprise as an economic institution whose purpose was, through an administrative structure, to control and co-ordinate a collection of activities undertaken by the various organisational units which made it up (Chandler, 1962; Penrose, 1959. For a critical examination, see Cowling and Sugden, 1987).

At the same time, there is substantial agreement on the identification of a sort of transition from an internationalised economy to a *global economy* - a term still widely used today without any precise definition - in the course of the 1970s. This period saw, on the one hand, the dismantling of the system of control of national economies laid down in the Bretton Woods agreement, followed by the broadening of the GATT regulations and the establishment of numerous free trade areas. On the other hand, it saw the strengthening and extension of the sphere of influence of the multinational corporations was accompanied by a profound reorganisation and restructuring of their activities.

More in particular, and with a considerable acceleration in the 1980s, the phenomenon of globalisation (Figure 2.1) has been seen in the increase in the trade of goods and services of all kinds[1], involving developing countries to an increasing degree[2]. A significant share of this trade is now between local branches or subsidiaries of multinational companies, which in the 1980s significantly accelerated what Michalet (1985) defined as 'delocalised production'. A progressively more

Internationalisation	Globalisation
• Rapid growth in international trade between industrialised countries	• More rapid growth of international trade between developing or newly industrialised countries
• Limited bilateral or multilateral agreements	• Multiplication of 'effective' free trade areas, broadening of international free trade rules and increase in the powers of world bodies (GATT, IMF etc.)
• Raw materials imported from only a few countries	• Raw materials or semi-finished products from almost all countries
• Trade above all in goods	• More rapid growth in trade of services than goods
• International investments almost exclusively in industrialised countries	• International investments in almost all countries
• Multinational corporations in a limited number of countries	• Multinational corporations in a very large number of countries (world companies)
• International trade in patents and human resources between branches of multi-nationals	• Multiplication of trade, specialised human resources, patents and scientific and technological information
• Technologies mainly national	• International technology
• Relatively slow evolution of product life cycles	• Acceleration of product life cycles following innovation and greater multinational competition

Figure 2.1 Some elements that differentiate the concept of globalisation from the internationalisation of the economy

important share of trade has also been channelled through newly formed alliances and agreements between companies in, for example, the automobile, computer and fine chemicals industries (OECD, 1992).

Finally, in the context of a growing increase in foreign investment from Japan and the European Union (previously, the dominant part came from the United States), the rising share represented by the services sector[3] explains the formation of large multinational corporations in finance and transport. In this scenario, it is useful to note how a large part of trade concerns the transfer of patents or rights for new products and new production processes. This has thus speeded up the adoption by companies of technologies of pluri-national origin, in addition to the rapid evolution of international flows of economic, technological, political and cultural information (Dosi, Pavitt and Soete, 1990; Harvey, 1989). It is on the basis of these structuring factors that, as is well known, Marshall McCluhan constructed his 'global village' scenario.

With the beginning of the 1990s, the globalisation process has continued to exercise its influence: while foreign investment has declined, there has instead been a more marked trend towards regional integration[4] and the formation of inter-company networks (in financial activities, R&D, production and commercialisation). This leads in effect to the definition of globalisation in terms of a form of internationalised production in which the activities generating value possessed, controlled and managed by companies are distributed in a number of markets. A growing share of value and wealth is thus produced and distributed through a complex 'spectrum of processes and relations' which integrate national economies to an unprecedented degree (TEP Report, 1992, p. 237)[5]

Global dimensions in the interpretation of economic and industrial phenomena

So far, I have presented some descriptive elements of globalisation that are sufficiently well known but that are in any case indispensable for transferring attention to some elements that appear significant for the issue in question. From our point of view, there are four key dimensions of globalisation inseparable one from another, which give an idea of the irreversible changes that have occurred in *industrial organisation*, which can no longer be explained as contingent events or events limited to individual sectors or countries.

First, there is the global extension of the company's economic horizon. The operating environment of the company, given the progressive contraction of space and time, tends to identify itself increasingly with the world economy. In other words, the point of reference for economic behaviour is a *varied* range (in space) and *variable* range (in time) of resources, markets, technological know-how, less and less dependent on national and continental borders.

Second, there is the growing *pluralism of technologies* and the strategic role played in economic competition by the development of scientific and technological potential. What is crucial, however, is that in his trend the speed of development and the diffusion of new scientific solutions are *not* as important as the *pluralistic* and *diffusive* way (in many countries, in many research centres) in which the technological frontier is rolled forward (Gaffard, 1990; Patel and Pavitt, 1989; Dosi et al, 1988). The specific nature of the new information technologies introduces a growing degree of *flexibility* - and thus variety and variability - in planning, production and the organisation of companies' decision-making systems: this multi-centred nature of innovative processes makes joint planning necessary, realised through the involvement of many different actors.

Third, there is the change in the nature of the relations between supply and demand, expressed in the assertion of a growing *autonomy of market demand*, which thus becomes a factor increasingly less controllable by the company. The globalisation of consumption should not, in fact, be understood (as happened in the Fordist logic) purely in terms of the homogenisation of needs or as a trend towards the development of standardised and uniform products in various market segments. While it is true that various examples suggest that this latter aspect is still present in some areas of production (Donaghu and Barff, 1990), it is nonetheless evident how the development of a global market leads to a rise in the quality of needs, and in the variety and variability of the products and services demanded, intensifying the significance of *country-specific* socio-cultural and institutional factors. In industrial economics, the transformation of the market into a variable that is increasingly independent of company decisions has represented a considerable theoretical outcome, after the initial insights of Normann (1979) and Ouchi (1984). Despite the reservations that have been expressed[6], this thesis has shown itself essential to the rediscovery of the value of the *plurality of spaces of reference* as a test bench of the real efficiency of the global enterprise.

Finally, in the growing complexity of the economy, the key variable is *information* and the control of information: information and knowledge appear in fact as essential to (a) *organisational flexibility*, implying the transition towards organisational forms capable of responding to challenges from a world increasingly characterised by instability, shortening of the product life cycle and globalisation of economic relations (Butera, 1990), and (b) *technological flexibility*, in other words by the introduction of computerised information systems, communications systems and programmed automation.

Given these dimensions of globalisation, it follows that in relation to *strategic company behaviour* two important insights are opened up that lie at the base of the argument that will be developed in this chapter.

First, it is now imperative that the company 'personalises' more and more its own products with regard to the different contexts in which they are offered, reducing the potential for standardisation. At the same time, however, the development of a cluster of fundamental innovations (in telecommunications, microelectronics and IT) have made the 'new' complexity of the system easier and more economic to govern.

18

Taken together, these two aspects - technologies and fluid markets - determine, on the one hand, an extraordinary shortening of the product life cycle, with the consequence that economies of scale and scope can only be pursued through the *simultaneous geographical expansion of markets* (Vagaggini, 1990, p. 85); on the other hand, they engender great instability in industrial sectors and companies, making competition more aggressive, allowing the entry of new competitors and increasing the replacement products.

In these conditions, it is thus possible to claim that a strategy will succeed if the company is capable of opening up to global competition, in terms of both geography and 'business area'. This is a form of company evolution that clearly contradicts the search for rationality and planning typical of the industrial corporation. The 'classical' model presumed a form of evolution (and internationalisation) based on the direct expansion of the company into a number of markets (both input and output), reducing environmental variability to the point of making the centralised management of major corporation structures possible. The current conditions of globality, technological flexibility and interaction with demand, *no longer* allow companies in general to be *self-sufficient* in the face of the problems posed by complexity. They therefore have to turn to specific capacities located outside themselves and out of their control.

Network configurations, as organisational forms aimed at enhancing both the specific features of the company and the capabilities located outside it, stand out in this scenario. This is obviously not the place in which to get to the root of these organisational structures which, defined generically as networked, in reality imply very different typologies. It is enough to observe how current processes seem to underline some common traits, among which are: (1) *interdependence* between enterprises, so economic efficiency stems from a series of systemic effects and cannot be easily evaluated on the level of the simple unit present in the network; (2) the *variety* of contractual forms which regulate transactions within the systems; and (3) *cooperation-dependence* as a consequence of asymmetry in the roles and structure of the actors belonging to the system (Conti, 1993a, p.115). On this point, valuable contributions have been made by the Scandinavian economic school (Håkansson, 1989; Johanson and Mattson, 1987 and 1993) and above all by the IEFI of the Bocconi University of Milan (Camagni, 1989; Camagni and Gambarotto, 1988; Vaccà, 1986).

Second, it appears at the same time reductive to discuss relations between companies (especially major corporations) and nation states in terms of confrontation between two different theses: between the supporters of the thesis that the states play a secondary role in relation to major corporations in the allocation of economic and technological resources, on the one hand, and those who claim that national powers still possess considerable means of intervention in economic and technological development on the world scale, on the other. If both these positions reflect reality in part, the new fact is that between states and corporations a new dynamic of *alliance* has been initiated, thanks to which the nation state assumes the function of ensuring its strategic actors - i.e. multinational companies - the possibility of realising the globalisation of the national economy and deriving political and social legitimacy from this (Petrella, 1989; Cerny, 1991).

Faced with the integration and the intersectoral nature of technologies, the growing costs of R&D, the shortening of the product life cycle, and the shortage of highly specialised personnel, companies, in addition to demanding legislation and political initiatives that favour their freedom of action, ask the state for specific support initiatives and services, such as:(a) the financing of immaterial infrastructures (basic research, vocational training, the diffusion of scientific and technical information); (b) tax incentives for research and innovation activities; (c) the guarantee of a sufficiently stable industrial base; and (d) the necessary commercial, diplomatic and political support (Petrella, 1989 p. 17. See also Porter, 1990).

Seen in the light of these two issues, the process of globalisation appears not (or not only) as the strengthening and extension of the great Fordist corporation's tendency to suppress differences and develop global products for unified markets. The global economy and the trend towards globalisation is not, in fact, a phenomenon that is 'totalizing and homogenizing in nature'(Le Heron and Park, 1995, p. 4). On the contrary, it 'is accompanied by (and even defined by) historically-specific patterns of different levels of complexity which are products of structured stability and coherence' (ibid, p. 5). Accompanied by the de-standardisation of production, the development of national varieties and the greater complexity of products and markets, globalisation makes specific national (regional, company) features the foundation of competition between different entities, where *variety is the origin of the production of wealth and competitive advantage*.

In contrast to the internationalisation that marked the 'glorious thirty years' after the Second World War, the new forms of relations are not in fact expressed in the mere international expansion of the individual company, but in the creation of *value chains* (and a division of labour between companies) founded increasingly on agreements and cooperation. The network structure therefore expresses the relation and the valorisation of different local identities which, precisely because of their diversity, can integrate together and evolve in a global scenario.

In this light, the territorial dimension represents a priority instrument for the explanation of the present industrial dynamic[7]. If the local is not separated from the global but is part of it, then it follows that the phenomena of globalisation generate a new *dialectic between local and global*. This dialectic is the very heart of this chapter. However, before exploring this dialectic, it is necessary, first to shift attention to the concept of local.

The search for the 'local'

Apart from the numerous ambiguities that connote the concept of *local development* (Greffe, 1987), it is reasonable to claim that in it the specific factors of transformation are emphasised and attention is focussed on the *territoriality* of each context, understood as an 'unrepeatable' set of social and economic relations. The intellectual wealth that is hidden behind the varied and sometimes confused and contradictory attempts that have marked research around the phenomenon of the local is indisputable. In part, this debate has pivoted around *visions* of a different development to the mechanistic, evolutionistic and globalistic canons [8] which have dominated debate in the post-war period, and which is reported in the literature in terms of development 'from below'. I prefer to talk here about 'visions' of a different development, not so much to underline the utopian and 'neo-populist' character (Gore, 1984) of many of these proposals (Sachs, 1980; Friedmann and Weaver, 1979; Stöhr and Taylor, 1981; Raffestin and Bresso, 1979; Cunha, Greer-Wootten and Racine, 1982), but because they do not form an articulated and self-contained theory of development, representing approaches whose 'intellectual history' - as Friedmann (1992) puts it - is still to be written. From the theoretical standpoint, these attempts should, however, be seen in the broader debate that has had the merit of driving to the wall the globalistic 'tyranny' that has held cultural hegemony over the social sciences in past decades (it is sufficient to cite the paradigmatic value that two completely different milestones have had such as Parsons' *Social System* of 1951 and *One-Dimensional Man* by Marcuse of 1967).

Despite having common intellectual roots, this body of thought should not in any case be confused with research on *local development*, which is again varied and difficult to summarise briefly. In order to give the most synthetic picture possible, a useful step is to turn our attention to the empirical and contextual motivations that lie at its foundations. They allow us, in fact, to define, even if roughly, some well-defined lines of research.

In Mediterranean Europe, interest in local development began above all with the 'success stories' of numerous small enterprise areas (Costa, 1991; Courlet, 1991; Pyke, Becattini and Sengenbergerl, 1990; Garofoli, 1994). In this case, the literature on the systems of small enterprises based its reasoning from the outset on *non-market interdependences* (Storper, 1992) and more in particular on the way in which society structures itself locally and organises production (Sforzi, 1993). The motivations that have inspired research in Britain appear rather different. Here, the accent has been placed above all on the problems of 'revitalising' the local economy, on relations between central and local government (Hudson and Plumm, 1986; Cochrane, 1985; Duncan and Godwin, 1984) or again on the introduction and diffusion of micro-electronic and information technologies (Morgan and Sayer, 1986). It is characterised as a whole by the lack of a theoretical approach that internalises an essential concept like that of *externality*. In the United States, the ample debate on post-Fordism and 'flexible accumulation', although offering original theoretical proposals [9] is largely based on the extension of the problem of the districts' small enterprises. Despite the wealth and diversity of the perspectives adopted, this debate has in general suffered from the substantial socio-cultural uniformity in the country, where rarely have the importance and variety of socio-cultural factors in the formation of the capacity for production and competition been examined (Dertouzos, Lester and Solow, 1991). Finally, it is worth remembering the significant (and, in part, original) debate that developed in France on the 'local production systems' (Aydalot, 1984; Pecqueur,

1987), clearly inspired by systems theory and which opened more than one insight into mechanisms for overturning traditional spatial hierarchies.

These perspectives differ considerably, both from the point of view of the subject of analysis and their theoretical foundations, while numerous overlapping areas are not to be excluded. There is partially common ground in the fact that these different perspectives converge in bringing to the forefront the *local as an intermediary unit of analysis*; in other words as the force that structures industrial organisation which is absent in traditional economics. The local, in other terms, is established by the need to go beyond the opposition between the macro-economic level of the national systems, on one the hand, and the micro-economic level of the enterprise-actor, on the other.

Nevertheless, this (in many aspects arbitrary) breakdown into different approaches has a theoretical significance that will be examined below. The thesis that I want to put forward is that the large quantity of research into district systems is a possible starting point for a *profound revision of the criteria of economic and social analysis*. It starts from the emergent view of the industrial district as a local system in which the system is a socio-cultural microcosm that cannot be separated from specific sub-cultural political features, and in which the production system cannot be isolated from the historical and cultural foundations that made it possible[10] (Brusco, 1982; Becattini, 1990).

The problem is neither that of identifying the defining criteria or the empirical identification of these entities, nor that of discussing whether the district does or does not represent an advanced form (or a variant) of Fordism[11]. The first would create a reductionist theoretical framework; the second, would continue an endless debate that has already absorbed considerable energy. The district is not merely an empirical model of reference on the basis of which to go on and identify the 'local', and not even, consequently, a normative hypothesis. The thesis proposed here is, on the contrary, that *starting from the cultural elaboration of the district it is possible to perfect a set of theoretical and methodological instruments of a more general value*, i.e. that can be transferred to ideal-types of differing territorialised production systems. In this light, 'the industrial district, as a concrete case of local development, is but one of the organisational and institutional aspects that the enterprise can adopt in the search for forms of flexible production organisation' (Sforzi, 1995, p. 43).

As I shall try to demonstrate in the following sections of this chapter, it has been above all the debate around the district phenomenon that has allowed the bridging of the fairly clear-cut gap between social disciplines, and as such has led to analysis of economic and production phenomena by abstracting them from social, cultural and institutional components. In effect, this scientific logic no longer appears feasible in the conditions of globalisation and greater complexity of the contemporary world and its economy. The following paragraph is devoted to this discussion, in which I shall attempt to break the question down, purely for illustrative purposes, into different sections, in which empirical elements and theoretical assumptions are mixed. On this basis it will be possible to give greater substance to the issue of the dialectic between 'global' and 'local'.

Some links in the problem

'Weak bonds' and territorial embeddedness

The assumption of the socio-cultural context as a key variable in the organisation of production obviously does not mean denying the importance that economistic schemes have had for the understanding of the contemporary processes of industrial change (here, it is sufficient to recall the contribution given by the transaction costs model, including its applications in geography) (Scott, 1988a and 1988b; Williamson, 1991). However, the latter bring with them a 'theoretical trap'. As Grabher (1993) notes, by abstracting the enterprise from its social context, they hinder the explanation of the importance played by changed market and technology conditions and by informal relations, on which relations between actors in a competitive market are increasingly based.

The foundation of the explanation solely on 'strong bonds' (technical and market relationships), in addition to re-asserting in some way the neoclassical logic, leads to the separation of the study of the organisational aspects of enterprise from that of its cultural and institutional background. The consequence is that the specific features and complexity of different local and company situations are 'reduced to contingent manifestations of a universal paradigm of economic rationality' (Becattini, 1989, p.7).

The latter has been a connotation of *corporate geography* itself, centred on the processes of technical and economic concentration and on 'virtuous' (and, in the end, deterministic) relations which are woven between companies, parts of companies and space. The problem is not, however, that of joining Walker (1989) in decreeing a requiem for this analytical perspective (which in some way has managed to describe the Ford-Taylorist enterprise and the related criteria of economic efficiency), but of refounding a corporate geography that is feasible in contemporary conditions.

Corporation geography and transaction theory itself[12] were products of a scientific logic characterised by the separation of social disciplines and by the of positivist economists' 'mistrust' of certain conceptual categories and the so-called *weak bonds* (such as identity, atmosphere, communicative interaction, industrial culture etc.) which, while being difficult to quantify, became protagonists as a consequence of the growing interdependency of the system and the lowering of spatial and temporal barriers.

These weak bonds, which inspired the thesis of *enterprise embeddedness* are not incompatible with the previous ones but are, on the contrary, part of a single cognitive spiral (Granovetter, 1985; Granovetter and Swedberg, 1992; Powell, 1990; Zukin and DiMaggio, 1990). It is in fact from the merger of these two components - the economic-productive and the non-market - that the *local dimension* establishes itself as the pivot for the revision of the criteria of socio-economic analysis and, in industrial geography, as the basis of a non-deterministic analysis of the relation between enterprise and space.

With the introduction of socio-cultural and institutional components (as intangible factors specific to each context), one can prefigure a sort of logical passage 'from the localised enterprise to the embedded enterprise' (Varaldo, 1995) which involves multinational corporations themselves, even if differently compared to the classic district enterprises.

On this point, a generalisation is helpful in outlining the idea of local not on the basis of a rigid dichotomy between local (district) enterprises and non-local enterprises (multinationals, to simplify things), but between *different forms and intensities of embeddedness*. This embeddedness will assume different forms (Varaldo, 1995, p.28): (a) for the district enterprise, embeddedness will be *natural* (i.e. connected to its entire life cycle), *totalising* (involving the entire network of its articulations, knowledge and culture) and thus *dependent* on local external economies; (b) for the multinational corporation it will assume a *planned* character (i.e. deriving from a preliminary choice of location), *selective* (i.e. aimed at interacting locally in different ways from one context to another) and *interdependent*, i.e. intended to activate specific interaction capacities locally. In this framework, the *co-evolution of enterprise and (local) environment* - understood as a specific set of tangible and intangible conditions - is both a pre-condition for the development of the company and a factor of reproduction of the diversity and multi-dimensional aspects of the local.

Large and small enterprises: competitive convergences

This way of conceiving the interaction between enterprise and environment leads to a redefinition of the problem of the competitive dynamic both for large and small companies (or better, for *local systems* of small companies). The traditional contrast between these two actors seems, in fact, to lose its theoretical importance, even if they each maintain their own specific features.

The phenomenon of the 'double convergence' of the two enterprise structures, already understood by Sabel but not sufficiently well explained, seems to find a more mature theoretical systematisation starting out from the theses on environmental embeddedness. It is not in fact enough to state that the large oligopolistic enterprise has begun to take and implement organisational forms that were once the exclusive domain of small and medium enterprise (Sabel, 1989): this can be true in certain sectors but not in others, and in certain national situations but not in others.

Assuming instead that competitive and (innovative) advantage derives not only from the interaction with other local enterprises, but, and perhaps more, from the embeddedness in a specific socio-economic and institutional context, this becomes, within certain limits, a common acquisition for both actors. This theoretical passage from local enterprise systems to global enterprise systems is extremely important:

> what characterises the development of global or transnational companies is still their capacity to embed themselves in the values, culture and behaviour of a given country, even if profoundly different to the one of origin ... in

that the evolution of technical-production and company organisation processes cannot be appreciated if one does not identify the specific ways that translate the socio-cultural and institutional potentials of the contexts into true productive forms of capitalist development. (Vaccà, 1993, p. 10)

The territorial embeddedness of small enterprise is a consolidated fact (Julien and Marchesnay, 1990; Conti, 1993b; Storey, 1993; Conti and Julien, 1989). If extended to the transnational or global enterprise - which involves different socio-cultural contexts in its interactions - this assumes an importance which affects industrial economics itself. As we have seen, embeddedness is not an automatic mechanism which is enough on its own to produce positive outcomes. On the contrary, it does not annul but valorises the enterprise as an actor that expresses its own *strategic behaviour*. Growing environmental complexity induces both the small enterprise and the global corporation to search for different styles of behaviour (in time and space), aimed at constructing more effective interactive relations, both with other enterprises and with the relevant socio-cultural context: in global competition, in fact, 'diversity becomes the basis of competitive advantage and the instrument for the production of economic value' (Conti, Malecki and Oinas, 1995, p. 3-4).

Multicultural interaction

Just as the socio-economic environment is not merely the expression of an autonomous historical evolution, but dependent on the strategic behaviour of economic actors, the enterprise is no longer conceivable as a self-sufficient system which establishes hegemonical relations with the environment. It becomes, on the contrary, *one* of the actors between which a complex dialectic relationship is established.

The problem is not that of wondering whether the era of the great monopolies is over or not. Profound organisational transformations - in particular, the formation of networked structures - have indeed strengthened them and at the same time made them more complex (Dicken, 1994). In this framework, the most effective management of local resources occurs through *culturally decentralised* company organisation, in other words, capable of interacting - and co-evolving - with the local environment. While expansion of the transnational company (through alliances, mergers, deverticalisations etc.) is an evident fact, it is nonetheless often accompanied by the segmentation of the company itself into smaller-sized flexible and autonomous centres[13]. This goes hand in hand with the decentralisation of the functions of control over actors who come into the company's operational universe in any way.

It is not possible to manage a global enterprise through control. In reality, hierarchical control tends to become the last choice when all the others seem to have failed. (Ohmae, 1993, p. 16)

25

From the point of view of the (global) enterprise, these forms of cooperative interaction (which do not deny competitive collusion) are the origin of the differentiation in forms of company development. And it is the network which increases the planning capacity of the various actors taken individually and of the system as a whole, allowing both to govern a more extensive and varied field of action.

The idea of the culturally decentralised enterprise thus allows us to leave behind once and for all the optimising schemes according to which the world outside the enterprise is a uniform context from a cultural and institutional point of view. This leads to the linking up of two elements, the transnational corporation and the country system, which from the theoretical standpoint had been kept artificially separate. It reaffirms that globalisation does not suppress but enhances national diversities.

This obviously does not mean that different contexts (national and regional) possess identical capacities for 'negotiating' with enterprise or for asserting their own socio-cultural values[14]. From a methodological perspective there is a significant lesson: the thesis that the multinational corporation is able (and tends) to produce the dissolution of local production and socio-cultural systems is now countered by a much less deterministic vision.

Globalisation and small and medium sized enterprises

While the thesis that TNCs are a vital element in contemporary economies is indisputable[15], globalisation is at the same time a phenomenon that increasingly involves small and medium sized enterprise (SMEs), which tend to exhibit levels of internationalisation qualitatively not dissimilar to those of large corporations. The denial of this reality leads in fact to the archetypal large corporation being assigned a character of universality that contradicts the logics of evolution of the contemporary industrial system. First, if one reasons in terms of the value chain, internationalisation is not a concept that can be referred only to some dominant actors. On the contrary, it implies a division of labour on a global scale involving a much broader range of actors. Secondly, it would hide a significant part of industrial reality: while it is true that the formation and the dynamic of the networks of coordination and cooperation have been described on the basis of examples of large-scale corporations, it should not be forgotten that the SMEs still constitute the backbone of the economy of the industrialised countries (companies with fewer than 500 employees at times make up 99 per cent of the total companies in a country) (OECD, 1992). This is not the place to restate the conditions of advantage and/or weakness of these actors in the contemporary economy: it has been demonstrated that the SMEs, even if with lower investments, achieve a level of radical innovation much higher than that of large corporations (Acs and Audretsch, 1988), just as the level of competitiveness on specific market segments is very similar to that of larger firms. Finally, in the last ten years the rate of growth of exports has been higher in many countries (such as Denmark, Spain, France, Italy and Québec) for SMEs than for large corporations (OECD, 1994)[16].

It is at the same time indisputable that the universe of SMEs is extremely varied and implies differing competitive behaviour. In the same way, the processes of globalisation lead to further segmentation of the fabric of the SMEs, appearing as a real factor of dynamism for many of them.

But on which mechanisms and processes is the globalisation of this type of company based? Again on this point, some generalisations allow us to grasp important issues. Figure 2.2 (constructed on the basis of work by Julien, 1995) divides the universe of SMEs into different categories according to the final markets of the products (*market space*) and the market origin of resources including information (*operating space*). The arrows indicate some major evolving trends, dependent on different *strategic behaviour*, but all aimed at coping with globalisation[17.] The squares in Figure 2.2 correspond to the following categories of behaviour.

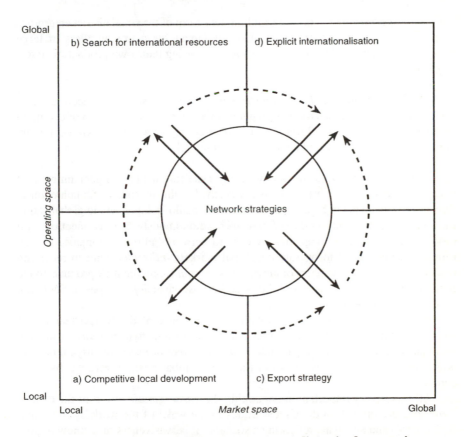

Figure 2.2 Globalisation and small and medium sized enterprises

Source: derived from Julien (1995)

(a)   *Local competitive development* (regional and national), in which SMEs make up the greater part. In this category are those companies obliged to improve their own competitive position in order to cope with increased competition from abroad. These companies can obviously operate in environments that are particularly dynamic and favourable to them (technological and financial support), and yet also follow major companies.

(b)   *Search for international resources* (especially information). While the product market continues to be local, the search for greater competitiveness drives some actors (especially in the services sector) systematically to turn to international sources, that they can then transfer to other locally-based actors. Long-term competitiveness requires an adequate flow of knowledge resources, even in traditional sectors. This is, therefore, a typical form of evolution of a district system.

(c)   *Export strategy*. This is an evolutionary form of the small district enterprise. It also represents a 'natural' form of development for many suppliers of large companies, which find the opportunities for increasing their own production assets in globalisation.

(d)   *Explicit internationalisation strategy*. This would seem to concern a rather limited number of SMEs and can be explained ideally as a possible form of evolution of some actors able to develop, for different reasons, their own economic, commercial, competitive and technological potential.

In the case of SMEs, we know in any case that a much higher number of companies operates on world markets *indirectly*. For this reason, a fifth behavioural dimension is included in Figure 2.2: this dimension allows for access to globalisation through *inclusion in international networks*, autonomously or in connection with other internationalised companies. Access to networks allows companies (which remain territorialised) to operate on global markets *indirectly*, both in receiving resources (from other members of the network) and in supplying their products to the market: this is a crucial aspect of the issue because virtually all types of SME can come under this heading.

Seen in this light, access to globalisation, because of the typical features of SMEs and their territorialised nature, makes the system dynamic, distinguishing between enterprises aiming at internationalisation more or less explicitly and others which draw advantage anyway from the 'new' global position reached by other actors.

All this is very important in debunking the *paradigmatic vision* of the district (widespread above all in the United States in the wake of the work by Piore and Sabel, 1984) and aimed at prefiguring a scenario of pervasiveness of the new regime of flexible accumulation. In this simplified vision, the enterprise as actor tends to disappear and be replaced by a new hybrid form, the *district-enterprise*, which behaves not too differently to a large corporation[18]. This scenario re-introduces very forcefully the enterprise's (and thus also the district enterprise's) role as actor and leads to a rather less deterministic vision of the district itself: as globalisation advances, the integrating forces within the districts lose bite and some dynamic actors, opening up to the outside, 'break the correspondence between the district and its operators. This transition does not necessarily "destroy" the local enterprise fabric, but *can favour its evolution through changeovers* ' (Rullani, 1994, emphasis added).

A global-local dialectic

The conceptual framework

Given these premises, the local organisation appears as a structural component of processes of a *global nature* which mark the development of the contemporary economy. By its very nature, the idea of local *is not* a dimensional concept but a way of conceiving of territory independently of scale, in the same way that the concept of global does not possess a dimensional character: it should not, in fact, be understood as 'extensive', but in reference to entities which are distributed in space through networks, nodes, flows, and whose borders are given by the expansionary capacity of the system (Giusti, 1991, p.144). This is the premise necessary to give weight to an idea of development *no longer* based on the assumption of a 'single' possible process of transformation, but on the plurality and autonomy of various levels of organisation and action.

If this rules out any conceptual polarisation, the acceptance of the dialectic between 'global' and 'local', even though it recognises the terms of the debate that have been reached, nevertheless demands its own scheme of explanation.

Faced with a problem of such complexity, and taking into account my own personal perplexity over the myths of the great theoretical categories and the problems of specialised languages, I shall proceed by breaking down the discussion into separate logical levels which can still be recomposed into a single overview. To this end, I shall assume three different approaches which in some way represent the principal theoretical proposals around which most of the international debate has hinged. With each I shall associate a level of discourse - respectively syntactic, semantic and pragmatic.

The latter, as we shall see, do not rule each other out although they do express different ways of interpreting the global-local dialectic and they are part of a single explanatory spiral. This is the only way in my view not to fall into the trap of excessively reducing complexity.

Significant logical foundations are found in Anglo-American industrial geography and reflect the intense debate around *flexible accumulation* which has occupied much of the research agenda since the mid 1980s. The perspective starts out from an explicit criticism of the thesis on the competitive advantage of the 'new' forms of specialised and flexible territorial organisation. It then re-asserts the protagonism of the large corporation and the major industrial and financial groups as the elements that bring about the accumulation and changes currently underway in the world economy. It is based on a dual foundation. Firstly, it is based on the fact that recent decades have been marked by an unprecedented wave of *concentration*, aimed at strategically repositioning the company in the changed technological, geo-political and competitive scenario. The new wave of mergers and acquisitions, although assuming new characteristics (inter-sectoral; touching both traditional sectors and technologically advanced sectors; involving SMEs) led in any case to the extension and the consolidation of the competitive position of large corporations on the international scale. Secondly, assuming production flexibility as an essential condition of competitiveness (Best, 1990)[19], these recent expansionary trends appear to assume different connotations than in the past. In particular, they seem to be characterised by experimentation with new organisational forms both inside companies and in inter-company relations.

In this light, the large corporation would seem to express an undoubted capacity for adaptation to the current economic conditions. Strategic alliances and relational structures where cooperation and competition exist together - in practice, the pursuit of network organisations - would be the demonstration of how large monopolies remain 'alive and well' (Martinelli and Schoenberger, 1991) and indeed strengthen their own hegemonical role in the system with different instruments compared to the past. The image of an industrial universe dominated by interlinked small enterprises would seem to fade and the geographical map of the world will not, in all likelihood, be a mosaic of new spaces of flexible production.

The enterprise, as a factor of globalisation and (in the search for a plurality and variety of conditions to pursue competitive advantage) as an actor in the dialectic between global and local, therefore expresses a strategy embodied in a *power relation* with the other fundamental competitive institution, the *nation state*.

Despite the present conditions of 'coalescence of national economies into continental blocks' (Gertler, 1992, p.270) and the rise of micro-regulatory institutions, the nation state is the context in which the institutional conditions for the creation of competitive advantage can be created explicitly. It is at the national level, in fact, that innovation policies are defined that fix the rules of the relations between industrial and financial capital, that the education and training system is structured, that the consumer market is structured. Gertler states, following Marcusen and Carlson:

So too do nation-states define distinct regimes of macroeconomic policy, which have the potential to: (a) produce differentiation in economic conditions (interest rates, unemployment, currency exchange rates etc.) between countries; (b) influence the basic conditions for production and the terms of competition for firms in each country, and (c) produce incentives for capital mobility within and between countries, as the national conditions to support manufacturing are made more or less favourable. (Gertler, 1992, p. 270)

Compared to a corporation looking for conditions of valorisation (which are localised in any case), the nation state represents the fundamental 'counterpart' for a dialectic of bargaining (which for obvious reasons excludes the local community) and is thus essential to the global-local dialectic.

In this light, the well known works by Dicken (1994) and Dicken and Thrift (1992) help to clarify this logic. By bringing the *entire production system*[20] into focus, they put at the centre of attention the problem of *organisation*, which finds in *networks of corporate governance* (Cowling and Sugden, 1987; Harrison, 1984; Miles and Snow, 1992; Wells and Cooke, 1991)[21] the instrument which answers to the conditions of growing environmental complexity and the search for flexibility. This extends the traditional concept of enterprise (the TNC assumes a coordinating function and has hazy boundaries) and the idea of the diversity of organisational forms is assumed[22], whose extension and configuration stem from the dialectic, which is a complex mixture of conflict and collaboration with the nation states (Dicken, 1994, p. 117), as global actors possessing specific competitive conditions and advantages. The nation states not only plays fundamental role as the country of origin of the corporation (in a TNC the home-base character - institutional and practices - invariably remains dominant (p. 118)), but also presents an opportunity for the corporation to take advantage of the regulatory differences. This interaction between enterprises and the state - both competitive institutions subject to global tensions - is thus essential to global-local relations. These are played out in *power relations*, as a rule excluding the lower spatial levels, which are 'relatively powerless, *except in a few very specific circumstances*' (Dicken, 1994, p. 123).

A nation state's boundaries therefore represent for the enterprise a 'major pressure' in their pursuit of a global strategy. It follows that the 'local' comes to be represented by that handy category widely used in economic theory, which does not admit any other behaviour than that of the system as a whole (apart from that of the individual actors). The 'local', as an intermediate entity (sub-national or regional) comes into the scheme of interpretation only as a residual category, 'currently being buffeted by the stormy seas of global economic changes' (Dicken and Thrift, 1992, p. 122-3). States and enterprises are assumed as 'totalising' categories, and as such give the scheme connotations of a sort of rationalistic determinism, which in this way cannot logically internalise the problem of territorial embeddedness.

31

Nevertheless, it is possible to give a sense to fundamental structural phenomena and processes which, *from the logical point of view*, appear independent of the characteristics of the single intermediate spaces in which production actually occurs. In the logical scheme presented here, this form of reasoning can therefore be depicted as a *syntactic* level[23], aimed at building some fundamental rules (invariants) that constitute and give meaning to the profound nature of the phenomena examined. The syntax of the global/local nexus is based therefore on a relation of power and, in strict terms, would need nothing else to be interpreted. Despite this, a syntax, if not completed by semantics and praxis, can lead to excessive simplification[24]. It is thus necessary to pay attention to a second logical level, which for reasons of clarity I shall compare to another particular type of approach now consolidated in the literature.

The neo-territorial approach: the dialectic of evolving relations

The logic according to which the pre-eminence of global processes linked to TNCs excludes local and regional systems - and therefore the rules and principles which define territorialities (Raffestin, 1986) - is hotly contested in geography, and also in regional economics, sociology, anthropology, political and management sciences. Starting from *regional economics*, a strong drive has begun to re-assess intermediate territorial spaces as places where processes of *collective learning* occur. The logic typical of 'globally structuring' processes operating at a world scale are accompanied by (and overlap with) different or *neo-territorial* logics, according to the definition given by Ratti (1991; 1992). They define the means by which *a region supplies adequate responses to the challenges from the outside (global) and in so doing produces variety and complexity, and thus globalisation.*

Bringing forward regional values, both in economic and socio-cultural terms, this perspective owes something to a debate that has shown itself particularly lively in the French-speaking world, supported by a considerable body of empirical research. Work on the *milieux innovateurs* (Aydalot and Keeble, 1988; Crevoisier and Maillat, 1991; Maillat and Perrin, 1992; Proulx, 1994; Perrin, 1989; Maillat, Quévit, and Senn, 1993; Camagni, 1991) and on local production systems (Pecqueur, 1987), it is not easy to summarise in a few pages. For this reason, I shall limit myself to identifying a cluster of partial reflections which, although relatively autonomous, reveal broader theoretical analyses.

This approach explicitly accepts the modern theory of industrial organisation, that describes the enterprise in terms of *functional relations* (and thus not only in terms of supply and demand in the strict sense), which occur within the framework of *three* distinct spaces of the firm's life (Ratti, 1991). The first two arouse no particular surprise.

(a)   *production space*, is determined by the relations of the functional and spatial division of labour involving different segments of production (the connection with the theory of transaction costs is explicit here): a company can produce autonomously or buy on the market, delocalising part of production according to the technological, economic and socio-cultural characteristics of the individual production segments and of the regions of production themselves.

(b)   *market space is,* defined by the market relations activated by the company, which can be interpreted spatially. They are obviously defined by the size, the intensity and the structural features of markets and their evolution.

The definition in spatial-functional terms of supply and demand had already represented, as is well known, a significant advance towards a dynamic theory of enterprise. However, in order to grasp the complex strategic and operational of companies open to dynamic international and global competition, it is necessary to introduce a third type of functional space:

(c)   *supporting* space, which embraces three different types of *'hors marché'* - non-market relations:

i.   *strategic-organisational relations,* determined by agreements between the company and its partners, which are put into practice in the exchange of information, in alliances and partial integrations;
ii.   *privileged relations,* between the company and the other actors that operate in the territorial space in question (public institutions, private or semi-public associations etc.);
iii.   *genetic-structural relations,* concerning relations in terms of the organisation of those factors of immaterial production (source of capital, network access to technological know-how, entrepreneurial culture and human capital) which constitute the *milieu* (or, by extension, the of Marshallian district).

This functional supporting space determines a significant part of the relations between supply and demand, i.e. between market space and production space. As most of these relations are integrated in 'non-market' relations (and largely precede the company's decision-making mechanisms), the supporting space is, in the final analysis, the basis of the success or failure of a given production choice. In this way, the functioning of the company is based on a complex set of often local relations which confer on the company itself its own *identity.* This marks the passage in some way from the *micro* level (of the individual company) to the *meso* level, in which the company becomes an *evolutive* (or co-evolutive) *actor* not separable logically from the dynamic process of environmental evolution. The context is explicitly systemic:

> when we speak of functional spaces, one must not understand simply the description of the territorial location of certain elements (attributes), their arrangement (pattern) and interconnection (structure) but, above all, the capacity to interpret these elements and structures in a process of

transformation, in turn oriented towards the needs of the entire system. (Ratti, 1991, p.67)

At this point we have come to the passage from the enterprise (or the global enterprise network), often non-identifiable territorially, to the system - the local network of economic and non-economic components that contextualises the supporting space. It is the (local) network, in fact, that makes the enterprise's supporting space an *external economy of organisation and information.*

The *evolutive* dimension of the (local) system thus transcends mere economic logics (such as the minimisation of costs) and takes on board broader values which assume meaning in the framework of the global/local dialectic. Faced with the global dimension of competition and the uncertainty deriving from the rapid introduction of technological change, the local system, 'as the territorialised materialisation of a set of structural relations' (Ratti, 1991, p. 84)[25] finds itself at the centre of the global/local dialectic. The characteristics and the identity of the local system are, from this standpoint, closely linked to the way in which the global/local relationship is realised *within the local system itself.* Indeed the capacities of local systems to restructure successfully (i.e. to preserve their competitive capacities in the global context) is positively correlated with their *system's identity* and, in particular, with the ability to change without losing (or reducing significantly) this identity: this is because a strong system identity allows a local system to respond to the impulses from the outside in a unitary fashion and yet flexibly, activating and valorising from case to case the different (more or less) latent 'classes of resources' according to the various dynamics emerging on the global scale.

The dynamic of the global-local relationship can thus be represented in terms of the (local) node-(global) network dynamic (Camagni, 1991; Conti, 1993a; Conti and Dematteis, 1995; Dematteis, 1994). It is made possible by the action of *actors* (enterprises connected to global networks and at the same time contextualised locally) which plays a dual function:

> On the one hand, they activate organisation in the global network, giving it internal coherence and aperture towards territorial systems. On the other hand, the same actor, representing the point of connection between different territorial systems, confers value on resources produced locally. In brief, the enterprise-actor is both an instrument of *closure* (i.e. it represents the organisational foundation both of the local system and the global network) and of *aperture*, as a vehicle connecting the system to the outside world. (Conti and Dematteis, 1995, p. 232)

Represented in this way, the local system is not a mere externality with regard to the enterprise, but a *collective operator* which can interact with the other levels (regional & local) through the intermediation of actors which 'dialogue' in the global network within a complementary framework of network and *milieu*. For example, 'when an enterprise chooses a partner with which to link up (in a network), it does not choose merely a single partner but a 'collective partner', given by the entire local culture and by the synergies of the partner's *milieu*' (Camagni, 1989, p. 227).

Communicative action as a network dynamic

In the logic described above, the socio-cultural components (which find a sort of synthesis in the concept of *milieu*) are seen as constituent elements of the competitiveness of the local system[26]. In this framework, the role of *knowledge* is assumed explicitly as fundamental to the network dynamic that develops both among local actors and between different local systems[27]. Nonetheless, full identification has not been reached of the mechanisms that can valorise a local *learning system* and make it competitive on the global level nor, on the other hand, of the mechanisms that determine (or inhibit) the territorialisation of knowledge. One of the great theoretical surprises of recent years has been that of getting to the roots of the dynamics of *communicative interaction*, interpreting them as fundamental in the global-local dialectic.

The theoretical foundations are found in systems theory (in particular in the idea of the observing self-referential system) and in the theories of communicative action proposed by Habermas (1986; 1988) and Apel (1989)[28]. In these theories, the actor's action is differentiated according to two perspectives - *strategic action*, involving an attitude of conflict, and *communicative action*, presuming instead an ideal of 'desire for accord' in interpersonal interaction. The second, in particular, is assumed as the dimension in which (a) cultural reproduction, (b) social integration and (c) socialisation occur, these being the basis of the process of the production of knowledge (Habermas, 1988, p.343)[29.]

These concepts constitute the logical foundation of a perspective that aims to explain the ways in which the economic and production system is integrated and is fed by its 'environmental background'. The latter, in providing labour, entrepreneurial and social culture, natural and immaterial factors and institutional organisation for production, is above all a territorially differentiated shell. Furthermore, not representing a mere container of given and unchangeable values and knowledge but evolving through *communicative (and synergic) interaction*, the local context constitutes a full-blown *laboratory* that produces accumulation of knowledge. It thus constitutes a critical resource in the development of contemporary capitalism and in the transition to the *information economy[30]*.

The basis of this theoretical scheme are the reflections of Nonaka (1995), a Japanese management scholar. Against the trends of the canons of dominant economic thought, he adopts the idea that knowledge *is produced and not only distributed*. In addition, the production of it derives not only from individual learning but is due to the *social action of organisations* (in this way, the author sets himself apart from Simon's problem-solving approach). Nonaka also accepts the thesis of the *dual* nature of knowledge, already advanced by Polanyi (1967), separating it into two distinct spheres from the point of view of content and language (and, as we shall see, their spatial dimension):

(a)   *explicit knowledge*, transferrable and explicable through a *code*. In other terms, it is exchanged through a scientific or technical language, by its very nature conventional, systematic, universal and/or shared.

(b)   *contextual knowledge* (or tacit, according to Polanyi), as the expression of the specific socio-cultural context that expresses it. Rooted in the action of the actors and in a given context, it maintains its meaning and validity solely in the (local) context that originated it.

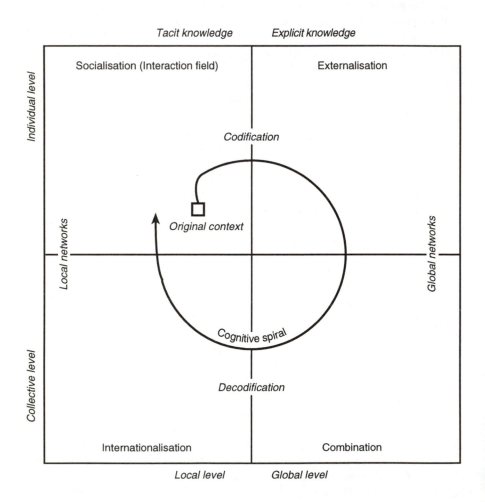

Figure 2.3 Process of knowledge conversion

Source: reworked from Nonaka (1995)

36

If a production system (whether global or local) of goods or services cannot be based on one of the two types of knowledge, but only on the reciprocal integration of both, the *local system* asserts itself as the key to the process of *knowledge conversion*. In it, the local system in fact has a dual function:

(a)   on the one hand, it connects the explicit and transferrable knowledge to production. In this conversion process, contextual knowledge, as the expression of local values and institutions, helps the local system to filter and transfer the codified knowledge according to its own needs. From the point of view of the meta-biology of knowledge, this immediately throws light on local embeddedness as an evolutive phenomenon connected to the aperture of the system, on the one hand, and on its self-referential nature, on the other. To tackle the external competitive and evolutive challenges, the system must, in fact, reduce the environmental complexity through a selection process based on a communicative code internal to the system. Furthermore, the system is itself self-referential, in other words, it is thus closed in on itself in its openness to the environment. This defines the response of the local system to the environmental (global) challenge.

(b)   on the other hand, it converts the contextual knowledge into explicit knowledge. The transferability from one context to another will certainly depend on the type of knowledge, but equally on the possibility of translating it into communicative languages through wHich the system is related to different places within global circuits of exchange of explicit knowledge.

A local system, understood as the place of *integration* of contextual knowledge and codified knowledge is not, therefore, a closed system, but a segment of a circuit (virtually global) of the learning and production of new knowledge:

> the production of knowledge is, in the end, a [localised] process: a process that occurs *in* predetermined places or *in the relation between* predetermined places ... The local dimension, in this interpretation becomes an essential element of the production of new knowledge. (Becattini and Rullani, 1994, p. 30)

Although reduced to the essentials, this scheme of reasoning allows some generalisations on the basis of which fuller conceptualisation of the global-local dialectic can be reached.

The functioning of the economic and production system would not be possible without the joint activation of (a) the *decontextualisation* of local knowledge (what allows the functioning and expansion of a production system based on the inter-regional and international division of labour) and (b) the *decoding* of explicit

37

knowledge, with the purpose of putting production, and, implicitly, the functioning and development of the enterprise into practice in concrete terms. In the evolution of the economic system from the industrial revolution onwards, this process of integration between the two cognitive spheres - global and local - has used different codes[31]: in the development of nineteenth century capitalism, the transfer of knowledge made ample use of the technological codes intrinsic to the machines which in Fordism were progressively replaced by organisational codes. The new emerging paradigm does not cancel the continuance of the previous codes, but adds other communicative ones to them, which become specific to and characterise a production system which substitutes uniformity (and the economies of scale of knowledge typical of Fordism) with *variety* and *variability*, as well as the interaction between different (local) contexts, as the factors of generation of competitive advantage. The local dimension must, in fact, transform itself through processes of relations, in an inter-local and virtually global dimension.

If Fordist rationalism was based on the decontextualisation and standardisation of knowledge, globalisation activates instead a spiral, thanks to which 'knowledge starts in contexts and returns to them' (Rullani, 1994, p. 60), in that only there can knowledge act directly as a productive force.

The problem of (global, local) *organisation*, and with it of *networks* (again at the global and local levels), thus becomes one of leadership, each one playing a function that is specific but not separable one from the other: 'The development of *global networks* endowed with a strong communicative capacity becomes an implicit necessity for the valorisation and the circulation of knowledge, just as, symmetrically, the strengthening of the contexts of origin and destination, as primary bases of knowledge, is - according to the scheme followed here - related to the growth in quantity and quality of the codified knowledge exchanged in the global networks' (Becattini and Rullani, 1994, p. 41).

This obviously does not mean that the world can be represented as a mosaic of local systems that are different but all equally competitive and capable of valorising their own accumulated knowledge. The socio-cultural systems capable of exercising a superior production capacity are those within which a continuous integration between the two spheres of knowledge is activated (i.e. those capable of activating their own substratum or *milieu*, by valorising it).

It is of course true that the multinational corporation can be seen as an agent which, assuming the global economy as a space without variety, devalues the contextual knowledge of the places where it operates (with the exception, perhaps, of its country of origin). There is no doubt about the fact that the dominance of global structures, now as in the past, is the origin of a global-local relationship that devalues (and destroys) the substratum and identity of places. From the theoretical point of view, this is not significant, however: the keystone to the question is represented here by the contextual dimension (local, but also national) as an essential variable in international competition (Albert, 1991; Lundval, 1990). In this light, the local system is an organised system endowed with a specific identity - as the expression of a set of values, knowledge and institutions. Its *evolution*, on the one hand, responds to external stimuli and, on the other, is the basis of the dialectic between the local system and global forces. It follows that, as the local system can be defined by the overlapping of the internal dynamic and the channels of access to globality, the global corporation itself can be understood only by 'giving theoretical weight to the specific nature of the places in which it is situated and not reduced merely to the network of connections in which codified knowledge flows' (Becattini and Rullani, 1994, p.37).

Conclusions: the global-local relationship as the culture of complexity

The scheme of reasoning followed in this chapter has allowed different levels of interpretation to be made logically autonomous, but at the same time not logically separable, each one bearing its own explanatory strategy. Despite being partial, this procedure was intended to illustrate the logic of the contemporary scientific method, characterised by theoretical multiplicity and the non-excludability of individual core concepts and levels of discourse. Upholding this thesis forces us, however, to broaden, in conclusion, the reasoning followed so far and to assume some categories typical of the debate around science.

Social research in general and economics in particular have provided in recent decades a mass of empirical results of great importance, which cannot be separated from profound conceptual revisions. It is, nonetheless, the epistemological area that remains the principal terrain of conflict. I believe that most of our efforts are in vain if we do not insert our reasoning in the framework of the broader debate on the nature of the transition - epistemological and ethical - underway in contemporary science. Only thus can *the problem of the global-local dialectic acquire full significance.*

To avoid getting into interminable epistemological discussions, we shall assume the metaphor that best illustrates the paradigmatic leap in progress in contemporary science. According to the felicitous intuition of von Foerster (1981), recent decades have seen the passage from a vision of reality understood as a *banal machine*, the expression of the mechanistic-structuralist paradigm, to that of the *non-banal machine*, the manifestation of the new paradigm that is being formed.

1.   The *banal machine* represents the world as an immutable set of elements and of relations between the elements, so the machine always gives the same response to the same impulse. It is its behaviour that demonstrates that the machine is (a) *analytically determinable*, because if one analyses it, its functioning can be understood; (b) *independent of history*, because it always behaves in the same way over time, laying bare its invariant nature; (c) *predictable*, as we can know what its behaviour will be over time (Vallega, 1994, p. 15).

39

The mechanistic ideal, aimed at imposing a 'simple' vision of the world, i.e. based on a linear logic of causality, could not but see space in the simplest and most abstract way possible: a passive support (a geometric one, in tHe most radical vision) in which the local territorial diversities were considered as irrelevant for the purposes of the explanation.

2.   The *non-banal machine* has different, even conflicting, properties. It possesses an *internal state* capable of determining what the machine does. Consequently, in the presence of the same impulse, repeated over time, the machine gives different responses. It follows that the machine (a) *cannot be determined analytically*, because one cannot understand how different responses are given to the same impulse, (b) *it is dependent on history*, because its behaviour changes over time, (c) its behaviour is *not predictable*.

This is obviously a metaphor, yet capable of illustrating facts that have undermined the foundations of conventional knowledge and the principles on which it was grounded (evidence, reductionism, causality, exhaustiveness)[32]. In great synthesis, the 'cultural revolution' underway in the contemporary sciences stems from a series of inter-connected phenomena:

1.   The establishment of *ecology as the foundation of a new epistemology*. I refer here not to ecology understood as a science that merely integrates different disciplines for the study of ecological systems, but as the bringer of a 'new', 'macroscopic' and holistic ethic and methodology for the study both of social systems and of phenomena of interaction between society and the environment. Perhaps not by chance in earlier pages, talking about co-evolution (between enterprise and environment), *milieu* and synergic interactions, an ecological language was used and the *evolutive* dimension was illustrated: in other words, the idea of irreversible changes that originate in the interaction between different systems.

Ecology, as an integrating category, is a systemic concept and thus expresses the explosion of systemism in the field of knowledge. To comprehend (from the Latin *comprehendo*, to embrace, envelop) the mechanisms and the functioning (of an ecosystem) means accepting its fundamental character, which is its systemic character. This marks the irreversible crisis of mechanistic determinism.

2.   The principle of reciprocal interaction between the whole and its parts is a foundation of the systems approach: it expresses the character of non-reducibility *between different disciplinary knowledge*, on the one hand, and, on the other, between the whole and the parts of any phenomenon. If assumed in isolation, a system can also be quantified, but in this way one fails to grasp the holistic quality, in that its dynamic and behaviour is inexplicable if isolated from the whole.

If different disciplines cannot be separated (although they are non-reducible), nor a hierarchy be made of them, no system can be separated from the whole of which it is part, nor from the parts which make it up. In this light, ecology and systemism represent the capacity to describe and understand the set of mechanisms and chains (*networks,* we would now say) which bind every system to others (from the infinitely small to the infinitely large): from this descends the heuristic character of a co-evolutive perspective and the *reciprocal non-reducibility between the local and global.* This reintroduces the global-local relation into every action of everyday life.

3.   This ecological, systemic and evolutive knowledge is the foundation of a true culture of complexity. It expresses the fundamental passage from the principle of *evidence* typical of mechanistic knowledge and its presumed scientific nature to the principle of *pertinence*, i.e. to the implicit or explicit intention of the researcher, endowed with his own cognitive structures.

The study of the phenomenon does not occur from a detached and abstract point of view, but from an angle imposed by the pursuit of one's own (the researcher's, the discipline's) ends. Complexity, as is well known, more than a property of reality, is a *way of describing it* that introduces elements of subjectivity into it.

In short, we find ourselves faced with a triangle whose corners - time, space, subjectivity - correspond to the three great founding ideas of contemporary science: (a) *time* expresses evolution, obviously not in terms of a trend towards equilibrium, but of irreversibility and unpredictability; (b) *space* refers to a composite object, in which the whole (global) cannot be broken down into parts, nor the parts be re-assembled into the whole; (c) *subjectivity*, finally, refers to the essential characteristic of the objects to be described, *whose property stems from the value attributed to them by the autonomous action of the actors.*

It follows that space (or better, spaces) has characteristics that do *not* stem from the 'general laws' of the (global) economy, but from its own organisational principles (in particular, the autonomous action of local actors). And here the geographical perspective comes decidedly into play:

> space is differentiated also and above all because of the long-term natural and historical processes and because of a time that is *neither reversible nor linear* in that it is an expression of unpredictable historical contingencies. (Dematteis, 1995, p. 4)

These contingencies constitute the substratum on which the organisational component of the territorially embedded actors can play, transforming local potential into values that can be exported in the global circuits.

41

The local autonomy of individual phenomena reintroduces the idea of the *contextualisation of the action of the actors* (Suchman, 1990) and, what counts the most, leads to the reinterpretation of both global and local systems, in terms of *self-organisation*: i.e. in terms of systems that are open and at the same time closed operatively, between which, thanks to the logics of the functioning of complex and autopoietic systems, *structural coupling* is accomplished[33]. The first, the global, is a space of flows which tie together different and changing local systems: it is, therefore, a *transterritorial* system. The second is instead a territorially closed system rooted in a specific *milieu* and its main function is to produce goods and services. The geographical perspective is that of giving meaning to the actors' action, in other words to the *embeddedness* in the values and local environmental endowments, whose evolution cannot be separated from the global dynamic of the system.

Notes

1. The importance of trade differs from country to country. It is significant for some small industrialised countries, such as the Netherlands and Portugal (whose exports represent respectively 36.7 per cent and 50.3 per cent of GDP), while it is relatively modest for other countries which already have a vast domestic market, such as the United States and Japan (10.3 per cent and 7.5 per cent respectively). Figures given in Julien (1995).

2. While between 1961 and 1975 the value of trade between the industrialised countries had risen by 7.7 per cent per annum and by 3.8 per cent with the non-oil producing developing countries, this percentage was overturned between 1976 and 1990, with the former dropping to 5.1 per cent and the latter rising to 8.8 per cent per annum. Figures from the International Monetary Fund, *Perspectives de l'Economie Mondiale* 1990, Washington, Table 19.

3. This is a phenomenon of primary importance in explaining the dynamics of the contemporary world economy. Deregulation and the interrelation *of* financial markets have contributed to this, as is well known.

4. In the meantime, much international trade continues to escape the GATT regulations, even though customs duties have fallen since 1945 from 40 per cent to 4 per cent. This can be explained by the multiplication of free trade agreements, by the fact that states adopt multiple measures to block imports and by the fact that trade is increasingly adjusted according to exchange rates defined by bodies that operate in parallel with the GATT (Petrella, 1989).

5. According to the Worldwatch Institute, growth in global economic output during the 1980s was greater than that during the several thousand years from the beginning of civilisation until 1950. In that decade, the growth in the world production of goods and services was indeed spectacular: it grew by $3.5 trillion between 1980 and 1990. Internal trade in goods increased in turn by 4 per cent per annum (Carley and Cristies, 1992, p. 101).

6. This thesis is challenged by, among others, Walker (1989), who sees the disarray of the mass market not as a cause but as an effect of changes in production, trying to cope with stagnation in Fordist production methods.

7.   There is now substantial agreement on this, and the wealth and variety of contributions to the literature in recent years have proved this amply. The lesson of Japanese and Korean capitalism (Dore, 1988; Dicken, 1992) and the one drawn from the experience of the Italian districts have been essential from this point of view, on a par with the debate in the United States on 'new industrial spaces' and on 'flexible accumulation' (see, among others, Gertler, 1992) and the French theoretical proposal of the *economie de la réulation* (Boyer, 1986).

8.   Also covered here is the broad path of post-War regional theory, impregnated with strong functionalist inputs, which joined with the evolutionist approaches that gave priority to the temporal-cyclical dimension of development over the territorial one.

9.   It is well known how one of the most well-trodden paths attempted to combine the framework of the regulationist school with other elements, in particular the emphasis on transaction costs.

10.   On the basis, the definition of the concept of *complete production cycle* was defined (Becattini and Rullani, 1994), referring to a process capable of reproducing not only the economic inputs in the strict sense, but also the social and institutional assumptions of production. The image that emerges is undoubtedly more convincing than that proposed by Dicken (1994).

11.   As it has been presented in the international literature, the figure of the district - and especially the *first generation* district - has rightly been accused of being the 'new orthodoxy' (Lipietz, 1992). The debate on second generation districts is now in full swing. In it has been noted the reduction in the internal forces capable of preserving a local nature (in the strict sense) to local companies, shifting attention to the capacity to open up to the outside, even if this can upset the original, consolidated network of relations, habits and interests: nonetheless, the initial fact, that the district represents a socio-economic category (i.e. a 'hybrid' in disciplinary terms) is maintained.

12.   In the so-called transactional paradigm, even transition towards post-Fordism is seen as descending from the 'objective laws' of capitalist development (i.e. immanent in the structure of the mode of production), assuming social phenomena as 'complementary'.

13.   Experience demonstrates how many multinational companies tend to acquire human and managerial resources capable of interrelating effectively with the local cultural environment. The relative autonomy of

the territorially rooted segments also reduces the constrains exhorted by bureaucratic centralisation: for example, Asea Brown Boveri, one of the world's major corporations, has reduced the central general co-ordination staff in Zurich from 4,000 to 200 people and has organised itself into a galaxy of 1,200 independent production centres. A not very different case is represented by the autonomy policy, including in strategy, conferred by General Motors to its own European affiliates (seeVaccà, 1993).

14. These conditions of 'weakness' of the local socio-cultural fabrics do not necessarily result in advantages for the enterprises, but they can also have effects in reducing their competitive capacity (see Brusco, 1994).

15. TNCs in particular, (a) dominate international trade and control almost all the trade in commodities such as bauxite, timber, copper, coffee and tea; (b) dominate direct foreign investment; (c) are a key source for technology transfer to developing countries; (d) are responsible for most of the technological innovations, including environmental protection technologies.

16. On the basis of the figures available, it is impossible to give an overall picture of the OECD countries. It is also important to note the share of indirect exports, due to the fact that the SMEs are often situated in value chains oriented to international trade.

17. A systematic explanation of this scheme is found in Julien (1995).

18. As is well known, this logic has led some authors to maintain that the SMEs' local networks are inevitably more flexible than the vertically integrated large corporation, thus putting themselves forward as competitive organisational forms in relation to the latter. The substitution of the enterprise by the district as the source of the generation of value and competitive advantage has led to the assumption of the district 'as a variation of Fordism: the difference was given by the organisational principle - the hierarchical power of the large corporation, the market organised in the industrial district - but with a single goal: the achievement of corporation, the market organised in the industrial district - but with a single goal: the achievement of economies of specialisation and scale connected with mass production' (Rullani, 1994).

19. This does not mean that there are no discordant interpretations, such as those that infer 'surprising rigidities' (Sayer, 1989) of corporate structures compared to (supposed) organisational changes.

20. This adds other fundamental components (technological progress, co-ordinative, regulatory, financial functions and labour processes) to the *production chain* (a concept substantially similar to that of *value chain*).

21. Nonetheless, Dicken and Thrift warn again 'falling into the trap of making hasty universal generalisations from a limited number of empirical cases' (Dicken and Thrift, 1992, p. 286).

22. At the same time, they are the product of a 'complex historical process of embedding which involves an interaction between the specific cognitive, cultural, social, political and economic characteristics of a firm's home territory' (Dicken and Thrift, 1992, p. 287).

23. Here, as a linguistic analogy, I use the concept introduced by Chomsky, 1981.

24. The image of the system that can be inferred is in fact that of an intensification, even if different to the past, of the hierarchical and asymmetrical relations in space. This is a consequence of the fact that only a few spaces can reach high levels of integrations, thanks to the concentration within them of international investments.

25. A *milieu*, like a trans-territorial network, plays two general functions: (a) it guarantees the static efficiency of the entrepreneurial system, by means of reduction of production costs and of transaction and co-ordination costs, (b) guarantees the dynamic efficiency of the entrepreneurial system, by means of the reduction of uncertainty in innovative processes (through information sharing, screening, transcoding and control), and establishment of the relational basis for (collective) learning processes and for construction of specific resources/competencies (Camagni, 1995, p. 198-199).

26. This allows, among other things, a reinterpretation of the concept of the *complete production system*, i.e. inclusive of all the activities (values, knowledge, institutions) necessary for the production of the material and human pre-requisites of production itself.

27. The introduction of the cognitive dimension no longer as a residual value is on the agenda in economic and management theory (Peters, 1992; Hayes, Wheelwright and Clark, 1988) and in part also in geography (Capello, 1994; Hall and Preston, 1983; Hepworth, 1989). Despite this, the current debate is marked by the contradiction of turning increasingly often to the

cognitive dimension while not having a complete theory of knowledge to hand.

28.   They develop, in their turn, the well known thesis of Schutz and Lackmann's 'vital world' (1974).

29.   These three processes occur through *abstract and specialised communnication media*. Within the economy and the political dimension of society these means are money and power, which allow the participants in social interaction to assume an 'external' attitude. On this question, see Giddens (1991), Geertz (1983), Lash and Urry (1993), Strange (1991).

30.   In this perspective, the thesis which states that knowledge is the decisive factor in determining development, rather than the accumulation of capital, is fundamental. Organisational changes must therefore be interpreted in the following manner: on the one hand, the problem of *reducing the costs of producing knowledge* definitively puts to one side the Fordist model. On the other hand, they respond to the need to *valorise this knowledge*, widening the field of application as much as possible (Itami, 1987; Morgan, 1988.

31.   'This integration can be achieved through: (a) *technological codes*, if the knowledge is incorporated in goods (materials, machines, components, finished products) and transferred with them; (b) *organisational codes*, if the knowledge flows from one place to another within the same organisation (such as a large corporation) thanks to the elements of homogeneity guaranteed by sharing the same hierarchical power and the same organisational culture; (c) *communicative codes,* if the knowledge is transmitted through communicative interaction, mediated by common languages and shared standards' (Becattini and Rullani, 1994, pp. 37-38).

32.   I refer here to the contrast, explained by Le Moigne (1990), between the precepts of the 'classical' (or Cartesian) modern scientific method and the precepts of the 'new' science taking shape. The principle of *evidence* is opposed to that of *pertinenece, reductionism* to *globalism, causality* to *teleology*, and *exhaustiveness* to *aggregativity.*

33.   According to the theory of autonomous systems (Wiener, 1965; Atlan,1979; von Foerster, 1981; Maturana and Varela, 1980 and 1985), the system is seen in a continuous state of dynamism, where the relations of reciprocal interaction with the environment (with other active systems) are defined in terms of *structural coupling.* This coupling occurs when the system, organisationally, closed, selects the perturbations from the outside,

modifying its own structure continuously. This process of adaptation which guarantees the invariance of the organisation is *autopoietic*, i.e. a circular process that reproduces the elements and the relations between the elements (i.e. the structure), modifying them. Paraphrasing Piaget, the concept of aperture therefore refers to the structure, and that of closure to organisation, which remains constant precisely because of the transformation of its components.

References

Acs, J.Z. and Audretsch, D.B. (1988), 'Innovation in large and small firms: an empirical analysis', *American Economic Review*, vol. 78, no. 4, pp. 678-90.

Albert, M. (1991), *Capitalisme contre capitalisme*, Editions du Seuil, Paris

Apel, K.O. (1989), *Etica del discorso come etica della responsabilità e il problema della razionalità economica*, Banca Toscana, Studi e Informazioni, Quaderno 24, Firenze.

Atlan, H. (1979), *Entre le cristal et la fumée*, Seuil, Paris.

Aydalot, Ph. (1984), *Crise et espace*, Economica, Paris.

Aydalot, Ph. and Keeble, D. (eds) (1988), *High technology Industry and Innovative Environments,*Routledge, London.

Becattini, G. (1990), 'The Marshallian industrial district as a socio-economic notion', in Pyke, F.,Becattini, G. and Sengenberger, W. (eds), *Industrial Districts and Inter-Firm Cooperation in Italy*, International Labour Organisation, International Institute for Labour Studies, Geneva.

Becattini, G. (ed.) (1989), *Modelli locali di sviluppo*, Il Mulino, Bologna.

Becattini, G. and Rullani, E. (1994), 'Sistema locale e mercato globale', in *Economia e politica industriale*, vol .80, pp. 25-48.

Best, M.H. (1990), *The New Competition: Institutions of Industrial Restructuring*, Polity Press, Cambridge.

Boyer, R. (1986), *La Théorie de la Régulation. Una Analyse Critique*, La Decouverte, Paris.

Brusco, S. (1982), 'The Emilian model: productive decentralisation and social-integration', *Cambridge Journal of Economics*, vol. 6, pp. 167-84.

Brusco, S. (1994), 'Sistemi globali e sistemi locali', *Economia e politica industriale*, vol. 84, pp. 63-76.

Butera, F. (1990), *Il castello e la rete: impresa, organizzazioni e professioni nell'Europa degli anni 90*, Angeli, Milano.

Camagni, R. (1989), 'Technological change, uncertainty and innovation networks: towards a dynamic theory of economic space', in Boyce, D., Nijkamp, P. and Shefer, D. (eds), *Regional Science: Retrospect and Prospect*, Springer Verlag, Heidelberg, pp. 211-49.

Camagni, R. (1995), 'Global networks and local milieu: towards a theory of economic space', in Conti, S., Malecki, E.J. and Oinas. P. (eds), *The Industrial Enterprise and Its Environment: Spatial Perspectives*, Avebury, Aldershot, pp. 195-214.

Camagni, R. (ed.) (1991), *Innovation Networks: Spatial Perspectives*, Belhaven Press, London.

Camagni, R. and Gambarotto, F. (1988), 'Gli accordi di cooperazione come nuove forme di sviluppo esterno delle imprese', *Economia e politica industriale*, vol. 58, pp. 93-138.

Capello. R. (ed.) (1994), *Spatial Economic Analysis of Telecommunications Networks Externalities*, Avebury, Aldershot.

Carley, M. and Christie, I. (1992), *Managing Sustainable Development*, Earthscan, London.

Cerny, P.G. (1991), 'The limits of deregulation: transnational interpretations and policy change', *European Journal of Political Research*, vol. 19, pp. 173-96.

Chandler,A.D. (1962), *Strategy and Structure: Chapters in the History of the American Industrial Enterprise*, The MIT Press, Cambridge MA.

Chomski, N. (1981), *Regole e rappresentazioni*, Il Saggiatore, Milano (Italian edition).

Conti, S. (1993a), 'The network perspective in industrial geography. Towards a model', *Geografiska Annaler. Series B*, vol. 75, no. 3, pp. 115-30.

Conti, S. (1993b), 'The University and the transfer of technological information to SMEs', *Small and Medium-sized Enterprises: Technology and Competitiveness*, OECD, Paris.

Conti, S. and Dematteis, G. (1995), 'Enterprises, systems and network dynamics: the challenge of complexity', in Conti, S., Malecki, E.J. and Oinas, P. (eds), *The Industrial Enterprise and Its Environment: Spatial Perspectives*, Avebury, Aldershot, pp. 217-41.

Conti, S. and Julien, P.A. (1989),'Le modèle italien: mythe ou réalité', *Revue Internationale PME*, vol. 2, no. 2-3, pp. 129-314.

Conti, S. Malecki, E.J. and Oinas, P. (1995), 'Introduction: rethinking the geography of enterprise', in Conti, S. Malecki, E.J. and Oinas, P. (eds), *The Industrial Enterprise and Its Environment. Spatial Perspectives*, Avebury, Aldershot, pp. 1-10.

Costa, M.T. (1991), 'La empresa espanola frente a la cooperacion internacional', in Velarde, J., Garcia Delgado, J.L. and Pedreno, A. (eds), *Apertura e Internacionalization de la Economia Espanola*, Coleccion Economistas, Madrid, pp. 397-416.

Courlet, C. (1991), 'El mercado comunitario del distrito industrial', *Papers de Seminari*, no. 33-34, Centre d'Estudis de Planificacio, Barcelona.

Cowling, K. and Sugden, R. (1987), 'Market exchange and the concept of a transnational corporation', *British Review of Economic Issues*, vol. 9, pp. 57-68.

Cunha, A., Greer-Wootten, B. and Racine, J.B. (1982), 'Le concept d'écodéveloppement et la pratique des géographes', in *Terrain vagues et terres promises*, P.U.F., Paris, pp. 7-126.

Dematteis, G. (1994), 'Possibilità e limiti dello sviluppo locale', *Sviluppo locale*, vol. 1, no. 1, pp. 10-30.

Dematteis, G. (1995), *Sistemi locali e reti globali*, IRIS, Incontri Pratesi su Lo sviluppo locale, Artimino, September.

Dertouzos, M.I., Lester, R.K. and Solow, R.M. (1991), *Made in America*, Edizioni di Comunità, Milano (Italalian edition).

Dicken P. (1992) *Global Shifts: The Internationalization of Economic Activity*, 2nd edition, Guilford, New York.

Dicken, P. (1994), 'Global-Local Tension: Firms and States in the Global Space-Economy. The Roepke Lecture in Economic Geography', *Economic Geography*, vol. 20, pp. 101-28.

Dicken, P. and Thrift, N.(1992), 'The organization of production and the production of organization: Why business enterprises matter in the study of geographical industrialization', *Transactions of the Institute of British Geographers*, vol. 17, pp. 279-91.

Donaghu, M.T. and Barff, R. (1990), 'Nike just did it: international subcontracting, flexibility and athletic footwear production', *Regional Studies*, vol. 24, pp. 537-52.

Dore, R. (1988), *Taking Japan Seriously: A Confucian Perspective on Leading Economic Issues*, Stanford University Press, Stanford CA.

Dosi, G. et al (eds) (1988), *Technical Change and Economic Theory*, Pinter, London.

Dosi, G., Pavitt, K. and Soete, L. (1990), *The Economics of Technical Change and International Trade*, New York University, New York.

Duncan, S. and Godwin, M. (1984), *The Local State and Local Economic Policy: Why the Fuss?*, Working papers in Urban and Regional Studies, no. 40, University of Sussex.

Friedmann, J. (1992), *Empowerment. The Politics of Alternative Development*, Basil Blackwell, Cambridge.

Friedmann, J. and Weaver, C. (1979), *Territory and Function: The Evolution of Regional Planning*, Arnold, London.

Gaffard, J.L. (1990), *Economie industrielle et de l'innovation*, Dalloz, Paris.

Garofoli, G. (1994), 'New firms formation and regional development. the Italian case', *Regional Studies*, vol. 28, no. 4, pp. 381-93.

Geertz, C. (1983), *Local Knowledge: Further Essays in Interpretative Anthropology*, Basic Books, New York.

Gertler, M.S. (1992), 'Flexibility revisited: districts, nation-states, and the forces of production', *Transactions of the Institute of British Geographer, New Series,* vol. 17, pp. 259-78.

Giddens, A. (1991), *Modernity and Self-identity*, Polity Press, Cambridge.

Giusti, A. (1991), 'Locale, territorio, comunità, sviluppo. Appunti per un glossario', in Magnaghi, A. (ed.), *Il territorio dell'abitare*, Angeli, Milano, pp. 139-70.

Gore, C. (1984), *Regions in Question. Space, Development Theory and Regional Policy*, Methuen, London.

Grabher, G. (ed.) (1993), *The Embedded Firm: On the Socioeconomics of Industrial Networks*, Routledge, London.

Granovetter, N. (1985), 'Economic action and social structure: the problem of embeddedness', *American Journal of Sociology*, vol. 91, no. 3, pp. 481-510.

Granovetter, N. and Swedberg, R. (1992), *The Sociology of Economic Life*, Westview Press, Boulder CO.

Greffe, X. (1987), *Politique Economique. Programmes, Instruments, Perspectives*, Economica, Paris.

Habermas, J. (1986), *Teoria dell'agire comunicativo*, Il Mulino, Bologna (Italian edition).

Habermas, J. (1988), *Il discorso filosofico della modernità*, Laterza, Bari (Italian edition).

Håkansson, H. (1989), *Corporate Technological Behaviour, Cooperation and Networks*, Routledge, London.

Hall, P. and Preston, P. (1983), *The Carrier Wave. New Information Technology and the Geography of Innovation 1846-2003*, Hyman, London.

Harrison, B. (1994), *Lean and Means*, The Free Press, New York.

Harvey, D. (1989), *The Condition of Postmodernity*, Blackwell, Oxford.

Hayes, R.H., Wheelwright, S.C. and Clark, K.B. (1988), *Dynamic Manufacturing. Creating the Learning Organization*, The Free Press, New York.

Hepworth, M.E. (1989), *Geography of the Information Economy*, Belhaven, London.

Hudson, R. and Plumm, V. (1986), 'Deconcentration or decentralisation? Local government and the possibilities for local control of local economies', in Goldsmith, M. and Villadsen, S. (eds), *Urban Political Theory for the Management of Fiscal Stress*, Gower, London.

Itami, H. (1987), *Mobilizing Invisible Assets*, Harvard University Press, Cambridge, MA.

Johanson, J. and Mattsson, L.-G. (1987), 'Interorganizational relations in industrial systems: A network approach compared with the transaction cost approach', *International Studies of Management and Organization*, vol. 17, pp. 34-48.

Johanson, J. and Mattsson, L.G. (1993), 'Strategic adaptation of firms to the European single market - a network approach', in Mattsson, L.G. and Stynne, B. (eds), *Corporate and Industry Strategies for Europe*, Elsevier, Amsterdam, pp. 263-281.

Julien P.-A. and Marchesnay, M. (1990), 'Sur le dynamisme des petites entreprises dans les pays industrialisés: Vers un nouvel équilibre entre les petites et les grandes entreprises', *Piccola impresa*, vol. 3, pp. 3-21.

Julien, P.-A. (1995), 'Mondialisation des marchés et types de comportements de PME', *Entrepreneurship and Regional Development*, forthcoming.

Lash, S. and Urry, J. (1993), *Economies of Signs and Space: After Organized Capitalism*, Sage, London.

Le Heron, R. and Park, S.O. (1995), 'Introduction: geographies of globalization', in Le Heron, R. and Park, S.O. (eds), *The Asian Pacific Rim and Globalization*, Avebury, Aldershot, pp. 1-16.

Le Moigne, J.L. (1990), *La Théorie du Système Général. Théorie de la Modélisation*, 3rd edition, PUF, Paris.

Lipeitz, A. (1992), 'The local and the global: regional individuality or interregionalism?', *Transactions of the Institute of British Geographers, New Series,* vol. 17, pp. 8-18.

Lundval, B.A. (1990), *User-producer interactions and technological change*, Paper presented to the Conference on Technological Change, 'Technology, Employment, Productivity Program', OECD, Paris-La Villette, June.

Maillat, D. and Perrin, J.-C. (eds) (1992), *Entreprises Innovatrices et Développement Territorial*, GREMI-EDES, Neuchâtel.

Maillat, D., Quévit, M. and Senn, L. (eds) (1993), *Réseaux d'Innovation et Milieux Innovateurs: Un Pari pour le Développement Régional*, GREMI-EDES, Neuchâtel.

Martinelli, F. and Schoenberger, E. (1992), 'Oligopoly is alive and well: notes for a broader discussion of flexible accumulation', in Benko, G. and Dunford, M. (eds), *Industrial Change and Regional Development: The Transformation of New Industrial Spaces*, Belhaven Press, London, pp. 117-33.

Maturana, H. and Varela, F. (1980), *Autopoiesis and Cognition*, Reidel Dordrecht, Holland.

Maturana, H. and Varela, F. (1985), *The Tree of Knowledge*, New Science Library, Boston.

Michalet, C.A. (1985), *Le Capitalisme Mondial*, Presses Universitaires de France, Paris.

Miles, R.E. and Snow, C.C. (1992), 'Causes of failure in network organizations', *California Management Review*, vol. 34, pp. 53-72.

Morgan, G. (1988), *Riding the Waves of Change: Developing Managerial Competencies for a Turbouent World*, Jossey-Bay, San Francisco.

Morgan, K. and Sayer, A. (1986), 'The electronic industry and regional development in Britain', in Amin, A. and Goddard, J.B. (eds), *Technological Change, Industrial Restructuring and Regional Development*, Allen and Unwin, London.

Nonaka, I. (1995), *The Knowledge-Creating Company*, Oxford University Press, New York.

Normann, R. (1979), *Le Condizioni di Sviluppo delle Imprese*, Etas Libri, Milano (Italalian edition).

OECD (1992), *La Mondialisation Industrielle. Quatre Etudes de Cas: Pièces Automobiles, Produits Chimiques, Construvtion et Semi-conducteurs*, Paris.

OECD (1992), *Politique Industrielle dans les Pays de l'OCDE*, Revue annuelle, Paris.

OECD (1994), *Globalisation des Activités Economiques et Développement des PME*, sous la direction de M.F. Estimé, C.Hall et P.-A.Julien, Paris.

Ohmae, K. (1993), 'La logica globale', in Bleeke, J. and Ernst, D. (eds), *Collaborare per Competere*, Il Sole 24 Ore Libri, Milano.

Ouchi, W. (1984), *The M-form Society*, Addison Westlay, Reading.

Patel, P. and Pavitt, K. (1989), 'L'Europa sta perdendo la corsa tecnologica', in Benedetti, E. (ed.). *Mutazioni Tecnologiche e Condizionamenti Internazionali*, Angeli, Milano, pp. 91-136.

Pecqueur, B. (1987), *De l'Espace Fonctionnel à l'Espace Territoir*, Thèse d'Etat, Université de Grenoble.

Penrose, E.T. (1959), *The Theory of the Growth of the Firm*, Basil Blackwell, London.

Perrin, J.-C. (1989), *Milieux Innovateurs: Eléments de Théorie et Typologie*, Centre d'économie régionale, Université d'Aix-Marseille III, Aix-en-Provence.

Petrella, R. (1989), 'La mondialisation de la technologie et de l'économie', *Futuribles*, no. 135, pp. 3-26.

Piore, M.J. and Sabel, C. (1984), *The Second Industrial Divide*, Basic Books, New York.

Polanyi, M. (1967) *The Tacit Dimension*, Routledge and Kegan, London.

Porter, M.E. (1990), *The Competitive Advantage of Nations*, The Free Press, New York.

Powell, W.W. (1990), 'Neither markets nor hierarchies: network forms of organization', *Research in Organizational Behaviour*, vol. 12, pp. 295-336.

Prahalad, C.K. and Hamel, G. (1990), 'The core competencies of the corporations', *Harvard Business Review*, May-June.

Proulx, M.U. (1994), 'Milieux innovateurs: concepts et application', *Revue Internationale PME*, vol. 7, no. 1, pp. 63-84.

Pyke, F., Becattini, G. and Sengenberger, W. (eds) (1990), *Industrial Districts and Inter-firm Cooperation in Italy*, International Labour Organisation, International Institute for Labour Studies, Geneva.

Raffestin, C. and Bresso, M. (1979), *Travail, espace, pouvoir*, L'Age d'homme, Lausanne.

Ratti, R. (1991), 'Le rôle des synergies locales dans les processus spatiaux d'innovation', *Revue Internationale PME*, vol. 4, no. 3, pp. 77-94.

Ratti, R. (1992), *Innovation Technologique et Développement Régional*, Méta-Editions, Bellinzona.

Rullani, E. (1994), 'Il valore della conoscenza', *Economia e politica industriale*, vol. 82, pp. 47-73.

Rustin, M. (1986), 'Lessons of the London industrial change', *New Left Review*, no. 105, Jan-Febr.

Sabel, C.F. (1989), 'Flexible specialization and the re-emergence of regional economies', in Hirst, P. and Zeitlin, J. (eds), *Reversing Industrial Decline*, Berg, Oxford.

Sachs, J. (1980), *Stratégies de l'Ecodéveloppement*, Les Editions Ouvrières, Paris.

Sayer, A. (1989), 'Postfordism in question', *International Journal of Urban and Regional Research*, vol. 13, pp. 666-95.

Schutz, A. and Luckmann, T. (1974), *The Structures of the LifeWord*, Heinemann, London

Scott, A. (1988a), *New Industrial Spaces*, Pion, London.

Scott, A. (1988b), *Metropolis: From the Division of Labor to Spatial Forms*, University of California Press, Berkeley.

Sforzi, F. (1993), 'Il modello toscano: un'interpretazione alla luce delle recenti tendenze', in Leonardi, R. and Nanetti, R.Y. (eds), *Lo Sviluppo Economico Regionale dell'Economia Europea Integrata. Il Caso Toscano*, Marsilio, Venezia, pp. 115-49.

Sforzi, F. (1995), 'Sistemi locali di impresa e cambiamento industriale in Italia', *Geotema*, vol. 1, no. 2, pp. 40-55..

Stöhr, W.B. and Taylor, D.R.F. (eds) (1981), *Development from Above or from Below? The Dyalectic of Regional Planning in Developing Countries*, Wiley J. Chichester.

Storey, D. (1993), 'United Kingdom', *Small and Medium-sized Enterprises: Technology and Competitiveness*, OECD, Paris.

Storper, M. (1992), 'The limits to globalization: technology districts and international trade', *Economic Geography*, vol. 68, no. 1, pp 60-93.

Strange, S. (1991), 'An eclectic approach', in Murphy, C.N. and Tooze, R. (eds), *The New International Political Economy*, Reinner, Boulder, pp. 33-49.

TEP Report (Technology and Economic Programme) (1992), *The Technology and Economy. The Determining Relations*, OECD, Paris.

Vaccà, S. (1986), 'L'economia delle relazioni tra imprese: dall'espansione dimensionale allo sviluppo per reti esterne', *Economia e politica industriale*, no. 55, pp.3-41.

Vaccà, S. (1993), 'Grande impresa e concorrenza: tra passato e futuro', *Economia e politica industriale*, vol. 80, pp. 7-24.

Vagaggini, V. (1990), *Sistema Economico e Agire Territoriale*, Angeli, Milano.

Vallega, A. (1994), *Geopolitica e Sviluppo Sostenibile. Il Sistema Mondo del Secolo XXI*, Mursia, Milano.

Varaldo, R. (1995), 'Dall'impresa localizzata all'impresa radicata', *Economia Marche*, vol. 14, no. 1, pp.3-25.

von Foerster, H. (1981), *Observing Systems*, Intersystems Publications, Seaside CA.

Walker, R. (1989), 'A requiem for corporate geography', *Geografiska Annaler*, vol. 71B, no, 1, pp. 43-68.

Wells, P.E. and Cooke, P.N. (1991), 'The geography of international strategic alliances', *Environment and Planning A*, vol. 23, pp. 87-106.

Wiener, N. (1965), *Cybernetics*, MIT Press, Cambridge, MA.

Williamson, O.E. (1991), 'Comparative economic organization: the analysis of discrete structural alternatives', *Administrative Science Quarterly*, vol. 36, pp. 269-96.

Zukin, S. and DiMaggio, P. (1990), 'Introduction', in S. Zukin and P. DiMaggio (eds), *Structures of Capital: The Organization of the Economy*, Cambridge University Press, Cambridge, pp. 1-36.

3 Global-local interdependencies and conflicting spatialities: 'space' and 'place' in economic geography

Michael Taylor, Carol Ekinsmyth and
Simon Leonard

Introduction

Trying to understand 'globalisation' and the dialectical tension between apparently separate global and local economic processes is an increasingly important and increasingly complex task for economic geography as it addresses the issues of 'space', 'place' and 'territory' that it involves (Amin, 1997; Amin and Thrift, 1994; Martin, 1994; Dicken, 1994a, 1994b, Thrift, 1986). Underpinning much of this globalisation literature is an implicitly hierarchic territorial framework that recognises the 'local', the 'national' and the 'global' as separate, though related, spatial scales and spheres of economic organisation and action (with the supra-national 'regional bloc' emerging as a fourth scale). In the various interpretations of these scales, the 'global' is viewed as an arena of world-scale processes involving flows of dominance and transformation that modify global-local relationships (Swyngedouw, 1992). At the national scale, the nation state is seen quite differently and alternatively as being either:

- subordinate to global markets and being replaced as a locus of regulation and organisation (the end of the nation state); or
- transformed into the 'transnational state'; or
- being relatively unaffected and retaining its independence and economic authority (Martin, 1994; Crook et al, 1992; Hirst and Thompson, 1992, 1996).

The 'local', in contrast, is caricatured as comprising place-bound fixities of tradition and continuity - industrial places and spaces that can only counter and combat overwhelming global economic logics through tight local integration and through 'learning' networks.

57

What is emerging now, however, is an alternative interpretation of this global-local dialectic that is concerned first and foremost with connection and connectivity, with linkage and interdependence between territories. The beginnings of this interpretation are identifiable in Dicken's (1994b) assertion that the problematic of:

> ... the "global-local nexus" ... is far more than merely a question of the geographic scale at which economic processes occur. More fundamentally, it is a question of *where power lies*. (p.102, emphasis added)

For Massey (1993) this is a 'progressive sense of place' whereby a 'place' is defined not by its counterposition with something labelled 'outside' but is defined through the very nature of its evolving and changing linkage and connection to that 'outside'. As it has been expressed by Amin (1997) drawing on Massey (1993), Held (1995), and Giddens (1996), this interpretation of globalisation as linkage and connectivity:

> ... replaces a territorial ideal of the local, national and global as separate spheres of social organisation and action, by a relational understanding of each as a nexus of multiple and asymmetric interdependencies involving local and wider fields of influence. It is the resulting interconnectedness, multiplexity and hybridisation of social life at every level - spatial and organisational - that ... [is] ... perhaps the most distinctive aspect of contemporary globalisation. Viewed in this way, to think of the global as flows of dominance and transformation and the local as fixities of tradition and continuity is to miss the point, because it denies interaction between the two as well as the evolutionary logic of both. (Amin, 1997, p.9)

We would argue, however, that as part of this reinterpretation of globalisation, notions of 'place', 'space' and 'territory' need to be further unpacked and more explicitly elaborated to recognise the nature of the spatialities involved. Without this unpacking, the connectivity, linkage and interdependence recognised by Amin and others can be interpreted, quite inappropriately in our view, as connectivity between *places* when, more properly, that connectivity is between agencies *in places* and between agencies with their own idiosyncratic *spatialities* and *territorialities* - the path dependent spatial frames within which they operate (the 'where's' of their knowledge - their spatial baggage). Without this unpacking we are left with Harvey's (1989a, 1993) detailed and elaborate contention that competition (as an expression of connectivity and the exercise of power) can actually be between 'places', implicitly ascribing to 'places' the quality of 'agency' usually ascribed only to people, business enterprises, organisations and institutions.

Agency and conflicting spatialities

The purpose of this chapter is, therefore, to explore the proposition that an essential component of the unequal power relationships that lie at the heart of economic interactions and the exercise of power in a globalising world are the spatialities of the actors and agents involved. Quite explicitly we wish to emphasise the role of agency in its many and varied forms in the globalisation process. Agency itself can be either the idiosyncratic agency of individuals or the collective agency of business enterprises (both SMEs and TNCs), organisations and institutions. The case for viewing organisation as a form of collective agency has been argued for strongly by Clegg (1989):

> organization ... [as] ... a form of collective agency ... is [not] ... a second-rate form of agency compared to that of the problematic human subject. Where organization achieves agency it is an accomplishment, just as it is for the individual but more so, because it involves the stabilization of power relations across the organizational field of action, and thus between many subjectivities, rather than simply within one embodied locus of subjectivities (p.188).

We would suggest that these very different actors and agents bring equally different, idiosyncratic and dynamic spatialities to the unequal relationships that are central to their survival and reproduction and their differential integration into a globalising world economy. These spatialities [1] are unlikely to be harmonised either within individuals or even within and certainly not between collective agencies. Conflicting spatialities will be the norm. Thus, to begin to illuminate trends and tendencies in global-local connectivities and tensions within the globalising space economy it is necessary to refine these notions of conflicting spatialities.

We are concerned to reinforce the well-rehearsed argument that 'space' and 'place' are not simply a canvass onto which social and economic processes are painted and configured. We are also concerned to argue that 'places' are to all intents and purposes illusory, agreeing with Harvey (1993) that 'place' '...has to be one of the most multi-layered and multi-purpose words in our language' (p. 4). But, we also wish to go further and to stress that, because 'space', 'place' and spatiality are individually and socially constructed, their essence and core characteristic is *conflict*: the confusing, confounding and contradictory conflict between the multiple spatialities of individuals, families, enterprises, organisations and institutions.

Our argument is quite simple. Individuals have multiple, dynamic spatialites reflecting their short-term daily activities, their longer term family, gender, social and employment relationships, their past experiences and prejudices and their stage in the life course. Enterprises, organisations and institutions have similar multiple, path-dependent spatialities fashioned by their transactions structures and interconnectedness within economies and societies, and the multiplicity of places (local and global) where they operate. At the same time, they incorporate at least

59

some of the conflicting spatialities of the individuals that comprise them. People within organisations bring with them a shifting spatial baggage that will affect managerial decision making just as much as shop floor resistance. The same is true of regulation and policy formulation at the level of the nation state, especially when government is interpreted as a set of bureaucratic organisations, hierarchically organised and involving the exercise of power and resistance in much the same way as in business enterprises. Indeed, these are the conflicting spatialities that appear to make drawing boundaries around localities, regions and places arbitrary. They are the conflicts that:

- confound the definition of travel-to-work areas for different workforce groups (Coombes et al, 1988; Peck, 1989);
- reduce the time geographies of individuals to mechanical reportage (Channing Adams, 1995);
- strip much of the meaning from Gouldian mental maps (Gould and White, 1974);
- render the place allegiances of corporations indeterminate;
- leave ambiguous the roles of elites within communities and nation states alike; and
- leave regulation and governance structures stranded at the level of the nation state (at least partially) and now, increasingly, at the supra-national level of the trading bloc (Redclift and Benton, 1994).

It is important to stress that in making this argument, we are not concerned with the *spatialities of outcomes* - the topographies of results and consequences that are built on antecedent conditions through the exercise of power and the ploys of resistance. Explaining topographies of outcomes has been a preoccupation of large sections of economic geography from Hartshorne's description of the Northeast Manufacturing Belt in the USA to studies of linkages and transaction structures and to the 'cargo-cult' planning of science parks, for example. Too often these have been the studies that in an unquestioning positivist and empiricist tradition have used pattern as a full expression of process. Neither, it is equally important to stress, are we concerned with the *spatialities of process* - the explicitly geographical processes acted out across space that have been the elusive crock of gold at the end of many geographers' rainbows. Rather, we are concerned with the *spatialities within actions* of individual or collective agency - the spatialities intrinsic to actors' actions that are themselves a mirror and reflection of the information, the prejudices, the foibles and the traits of the agents that perform them. Spatiality in this sense is like time, it is neither outcome nor itself process.

There is then no simple geography to be read off from these multiple, overlapping and conflicting spatialities within actions. There is no 'global' without its constituent and constituting 'local'. 'Space', 'place' and the spatialities of agents, redolent with confusion, contradiction and conflict, are intrinsic to the relationships which drive the economies and societies of places in particular directions and link them uniquely into the global economy. Agents' spatialities are intimately bound into every decision, action and strategy that prompts or responds to change. Spatialities are thus part of the learned environment within which agents operate; they are information itself, and neither the context within which other information is assessed and assimilated nor a topography of outcomes. They are an essential and intrinsic element of the asymmetric linkages and connectivities between the global and the local and, as knowledge, they are integral to power asymmetries.

In the remaining sections of this chapter we develop, elaborate and illustrate these propositions on conflicting spatialities in the economic context of global-local connectivity and interrelationships from three separate but overlapping perspectives:

- strategic management and the agency of individuals;
- the business enterprise and collective agency; and
- local labour markets as places where the contradictions of the conflicting spatialities of individual and collective agency are most immediately apparent.

Drawing on the work of Polanyi (1957) and others, it is possible to interpret 'conflicting spatiality' as the spatial underpinning and spatial expression of the intertwining of 'reciprocal', 'redistributional' and 'market' forms of integration in societies. Equally, it can be argued that stabilising processes in these societies, embracing economic modes of social regulation (Jessop, 1990; Peck, 1996) together with rule making and the creation of meaning and membership (Callon, 1986) and real regulation through formal systems of law and governance (Clark, 1992), can create a temporary stabilisation of conflicting spatialities. This stabilisation would involve a temporarily tolerated level of inequality which would, in turn, produce a stable economic-cum-geographic outcome in a range of places - a sort of spatial fixity, even a new spatial division of labour (Massey, 1984). These issues will be explored and developed in the final sections of the chapter.

Strategic management, the individual and conflicting spatiality

As the role of agency (both individual and collective) has received increasing attention in the debate in economic geography on global-local interdependencies and the processes shaping topographies of production, consumption and restructuring, so the importance of strategy formulation, strategic decision making and strategic choice has begun to be recognised - linking structure, strategy and spatial outcomes both global and local. In this emerging debate there are suggestions that different and conflicting spatialities need to be recognised and appreciated before the geographical patterning of restructuring responses to globalisation processes can be understood.

In the emergent political economy approach in economic geography the debate on restructuring has attempted (both implicitly and explicitly) to link corporate strategy and decision making with corporate structure and spatial outcomes.

61

'Structural' pressures in the global economic environment have been interpreted as stimulating the formulation of purposive strategies by 'capital', leading to appropriate and inevitable restructuring with equally appropriate and inevitable local outcomes. This stream of reasoning is only too clear in the literature promoting influential theses on the New International Division of Labour, spatial divisions of labour, flexible accumulation and flexible specialisation, and networks and embeddedness that underpin geography's evolving interpretation of the capitalist project in which emphasis is placed squarely on strategic outcomes. It has been criticised as deterministic and, in management science terms, 'environmentally deterministic' (Whittington, 1988), in that an agent may choose from a wide range of available courses of action but the economic environment will ensure that only one outcome is attainable. In geography, Clark (1993) too has recognised components of this determinism in the geographical literature on corporate strategy and restructuring (embracing strategic choice and strategic decision-making) suggesting that much of it treats corporate strategy as '... disembodied autonomous 'forces' separable from identifiable agents' (p.24). He has, however, also been accused of perpetuating the same determinism by insisting on the intentionality of strategy in response to market competition (McGrath-Champ, 1995).

At its worst, therefore, the exploration of corporate strategy and strategic decision making in economic geography has focused on strategic outcomes rather than on strategic processes. 'Strategy', it has been maintained has become a chaotic concept that needs unpacking and, in particular, 'accumulation strategy' '... has become a mantra, a ritualistic, largely unexamined, thematic repetition' (McGrath-Champ, 1995, p. 14). From the perspective of the 'conflicting spatialities' hypothesis, it can be suggested that McGrath-Champ's criticism centres on 'spatiality' (essentially the separate and conflicting spatialities of individuals within a collective agency) having no role as process in current treatments of corporate strategy, only a role as outcome - the game board on which economic processes are played out.

The need to elaborate and deepen the concept of spatiality in relation to the economic-geographic dimensions of strategic management are evident in studies by Clark (1993, 1994) and Schoenberger (1994). Against the background of sunk costs, Clark (1994) has developed in detail the notion that the history and geography of a firm - its path dependence - will manifest itself in ways (structures) that limit a decision maker's scope for action. He has explored the relationship between structure, strategy, history and geography and concluded that '... history and geography matter, but their significance is mediated through the structure-strategy relationship' (p. 30). However, he also notes that firms in the same industries often have similar sunk costs, similar histories and similar geographies (i.e. similar path dependence) but adopt different strategies. He offers no explanation for this divergence and apparent inconsistency. From the point of view espoused here, this apparent inconsistency might be better understood as a product of the *spatialities within actions* of individual strategy makers that must necessarily be reconciled within the collective agency of a business enterprise if it is to remain commercially effective. What is missing from the analysis and needs to be investigated is the information bases and intrinsic spatialities of the individuals who make decisions within firms in order to arrive at reasons for differences outcomes: what is needed in an investigation the histories and geographies of key decision makers in business enterprises not just the histories and geographies of the enterprises themselves.

Schoenberger (1994) takes Clark's argument further. In an attempt to theorise the corporate strategist, she notes that corporations do not always respond appropriately to external changes that put pressure on them to restructure and that, in the face of evidence and information about what they should do, they still do not always act accordingly. She puts 'structural sources of industrial rigidity' (p. 435) to one side, and identifies an 'unexplained residual' (p. 436) when firms with seemingly similar structural rigidities do not act in the same way. This unexplained residual is ascribed to the people who devise corporate strategies, and Schoenberger argues that we '... need to understand something about [these people] ... as social agents in a particular place and time' (p. 436) in order to understand why they make the decisions they do. This is an important insight and we would argue that just as history and geography can '... conspire to limit the scope of firms' responses to new initiatives in the market' (Clark, 1994, p. 18) an individual decision maker's personal history and geography can conspire to modify or limit his or her thinking or problem solving capabilities and capacities. These personal histories and geographies are, we would venture, more than Schoenberger's (1994) 'trained incapacities' or 'occupational psychosis' or 'the rigid commitments that prevent the individual from seeing the world in a different way' (p. 441), they also embrace the evolving, multi-layered spatialities which individuals carry with them associated with their life experience and which are built into their very actions.

Strategic decision makers are social actors with deep social and cultural foundations that are, to a greater or lesser extent, individual and unique and change through their life courses. We argue, in common with the early Humanists, that place and space histories are intimately bound up with individuals' personal senses of self, identity, security and well-being (Tuan, 1975, 1977, 1982; Prohansky, 1978; Seamon, 1979; Buttimer and Seamon, 1980). In reasserting the importance of the individual in corporate decision making, we are asserting the importance of place and space as factors which are intimately bound into the personas of the individuals in question. We argue that the roles of place and space, and time-space histories, in the context of organisational decision-making *from the point of view of the individuals making those decisions* are substantial and untheorised. They become increasingly complex when they are added to the time-space histories of business enterprises and organisations, industries, countries and cultures. In essence, we contend that human social behaviour is made up of overlapping and interacting spatialities which may concur, conflict and even confound both within the individual and between individuals and groups in a broadened definition of agency.

In accordance with this interpretation, Schoenberger's (1994) innovation and vision does not perhaps go far enough in theorising the corporate strategist. Personal space-time histories must surely affect and be embedded in strategic decisions and be manifested in corporate spatial outcomes.

Top managers are often powerful agents in local communities, occupying positions perhaps on local TEC boards, Rotary and other service clubs, churches, or simply as members of local golf, tennis and other sporting clubs. These interests could very well lead to conflicts to do with place and space if, for example, a decision has to be made to close down a particular operation with consequent job losses in a place that the decision maker has some personal allegiance to. At the same time, through time-space compression afforded by advanced communications, actors can operate in and affect the daily work processes of places anywhere on the globe that they may never even have been to and of which their experience is entirely indirect. Their perceptions and images of those remote-to-them places are at one and the same time increasingly vital and increasingly prone to being totally inappropriate.

Middle managers, in contrast, might be the recipients of orders from headquarters in another place or country which they are required to put into operation in their 'home' place. In this kind of scenario, managers at different levels in corporate hierarchies can be expected to have differing spatialities, perceptions and allegiances to places, and there is significant potential for these spatialities to conflict. Middle managers in public and private bureaucracies have long been recognised as the most resistant to change. Indeed, we would contend that just as the exercise of power or the attempt to exercise power is bound up with personal and conflicting spatialities, so resistance too can be ascribed to personal spatialities that conflict with those of centres of control.

People with the greatest allegiance to the places in which they work and live are also those with the least power - the workers. These people tend to be the least mobile through their life courses and are thus likely to be working in places with which they have profound psychological bonds and beyond which they have no knowledge or experience. The opportunities for and likelihood of resistance to any form of change in these circumstances, as a result of conflicting spatialities between individuals and communities on the one hand and public and private bureaucracies on the other are, therefore, substantially magnified.

At this stage, the importance of conflicting spatialities within the *processes* and not just the *outcomes* of strategic decision making and strategic choice is all conjecture. What is needed is a bringing together of the literature from economics, management science, behavioural and humanistic geography, the cognitive sciences and social theory to more fully conceptualise the role of place in the make-up of individuals and thus the role of spatiality in corporate and business enterprise decision-making. We agree with Schoemaker (1993) when he maintains that to effect the melding of the organisation and cognitive sciences with economic science '[a]dditional synthesis and pluralism is needed to advance *integrated* theory development - as opposed to fractionalism and applied fractionalism - in the field of strategy' (p. 108). We wish to broaden this agenda to include the role of place-space histories or 'cognitive spatialities'(along with their inherent conflicts) in developing a fuller understanding of the decision-making processes of corporate and business enterprise strategists.

The business enterprises and conflicting spatiality

From the perspective of the business enterprise and economic restructuring, aspects of conflicting spatialities have always been evident in economic geography, but only partially. The tension between what might loosely be termed 'corporate' space and local communities (or 'places') was quite explicit in Massey's seminal work on spatial divisions of labour, but space and place were also put centre stage to transform competition *between business enterprises* into the *competition between places* that also lies at the core of Harvey's (1989a, 1993) analysis of the condition of post modernity.

Working within a somewhat different framework, Dicken (1994a,1994b) and Dicken et al (1994) have identified a different set of tensions between globalising and localising processes as they affect the dynamics of societies and economies, maintaining that '..dynamic power relationships between transnational corporations and nation-states is a critical determinant of local economic development' (1994a, p. 218). Here, two basic and conflicting spatialities are apparent. First, there are complex corporate spatialities, and Dicken cites the modes of cross border corporate control identified by Bartlett and Ghoshal (1989) together with Storper and Harrison's (1991) analyses of forms of production system governance to point to differential power between the functional rather than the purely geographical cores and peripheries of transnational corporations. These ideas are not new: they are well rehearsed in the management science literature, and especially in the work of the Aston Group in the late 1960s, and in the geography of enterprise literature (Taylor, 1987,1995). The role of the 'home' country is still ambiguous in this literature with Dicken (1994) disagreeing with Reich (1991) and Ohmae (1990) and maintaining that '... a TNC's domestic environment remains fundamentally important to how it operates, notwithstanding the global context of some firms' operations' (Dicken, 1994a, p. 234). Second, there is the very obvious spatiality of nation states, and nation states increasingly are seen as a pivotal players in the economic processes of globalisation and internationalisation. To Gertler (1992), they generate particular systems of commercial and industrial innovation, while to Christopherson (1993) and Clegg (1990) they also reflect different cultural values and create different climates of regulation and governance. Here then are the conflicting spatialities of the nation state and the corporation bringing implications and consequences for 'places': two geographies played out in 'places' - the same interplay identified by Corbridge et al (1994) in relation to globalising, internationalising money.

A counterpoise to this emphasis on the tension between nation states and the variable governance and control structures of TNCs, is the literature on flexibility, 'new industrial spaces', 'embeddedness' and 'institutional thickness' that to all intents and purposes gives primacy to 'place' and interprets interrelationships between enterprises in reciprocal rather than competitive terms. Network relationships are seen as spatially specific and collaborative and are modelled on the interpretations that have developed of industrial growth and change in such places as the Third Italy and Baden-Württemburg (Goodman et al, 1989; Sabel, 1989;

Lazerson, 1993). This particular model of transactional relationships is drawn largely from sociology, and relies heavily on the work of Powell (1990), emphasising strategic alliance, trust and reciprocity in the interrelationships that appear to weld SMEs into functionally effective combines. Here, institutional support and the socio-political foundation of economic activity is afforded a central and determining role - a new language and deeper understanding of the agglomeration and locational integration recognised by Florence (1948) and a reinvigoration of Marshallian industrial districts. At the same time this literature broadens the scope of spatiality by incorporating into the socio-political game of economic change the place-specific characteristics of a range of other entrepreneurial and institutional players.

But, taken from a business enterprise perspective, neither of these approaches identifies what, from the perspective of the argument proposed here, we would see as the full range of spatialities that come into conflict in any dynamic situation of socio-economic change. At least four sets of spatialities can be identified in the institutional, organisational, enterprise arena; what might be interpreted as a caricature of multi-layered conflicting spatialities. These sets comprise:

- *intra-enterprise spatialities* - between the unequal elements of multi-site enterprises including multinationals, transnationals (Dicken, 1992) and multi-domestic corporations (Porter, 1986), involving the different control and delegation systems of multi-divisional and holding company structures (so-called H-form and M-form enterprise structures), matrix structures, 'mother-daughter' structures (Franko, 1976) and 'reserve power' forms (Handy, 1994);
- *inter-enterprise spatialities* - between transnational corporations and small and medium sized enterprises (SMEs), for example, reflecting shifting bases of competition, collaboration, cooperation and collusion, and reflected in such arrangements as strategic alliances, joint ventures, subcontracting and franchising, and involving relationships of reciprocity, trust, cooptation and imposition;
- *the spatiality of the nation state* - which is said to be being 'hollowed out' through the forces of globalisation and internationalisation, but which can be interpreted as being quantitatively and qualitatively restructured and reformulated; and
- *the spatiality of the local state* - including local governance structures and local regulation;

Where the spatialities of the nation state and the local state should be cast in terms different to those used to depict intra- and inter-enterprise spatialities is a moot point. Sections of the management science literature would interpret governmental bureaucracies as just as much organisations as are private sector business enterprises (Pfeffer and Salancik, 1978). They are, by this interpretation, no more than another form of competing, controlling and complementary organisation contributing to the networks of inequality and unequal power relations within which business enterprises are embedded.

However, complicating and in some ways confounding these business enterprise and institutional spatialities are the personal spatialities of the people that operate, plan, drive and manage them. All these people bring to their employment (their day-to-day involvement in the purposive collective agency of business enterprises, organisations and institutions) their individual, familial, gender, ethnic and cultural biases and orientations that must necessarily impact on their decision-making and resistance to the exercise of power. Thus, the multi-layered, conflicting spatialities of enterprises and institutions can not be separated from the multi-layered, conflicting spatialities of the people that produce and reproduce them. In this way the conflicting spatialities of collective agency in the form of business enterprises and other forms of organisation merge with the conflicting spatialities involved in strategic management, strategic choice and business strategy formulation.

Local labour markets and conflicting spatiality

In contrast with the business enterprise and strategic decision making contexts where spatiality has had to be, in effect, 'discovered', local labour market studies have tended to begin with 'space' and 'place' which, we contend, has served at best to obscure but at worst to overwhelm the conflicting spatialities of individual and collective agency.

Local labour markets dominate peoples' everyday lives and local labour market theories tend to reflect this relationship and to be specifically space and place related. They give primacy to capital-labour relations in 'place' over capital-capital and capital-labour relations with different spatialities. In this sense, theories of the local labour market are arguably partial and exclusive of the body of literature which centres upon the firm, business enterprise and transnational corporation where 'space' rather than 'place' has been the central geographical focus.

The regulation theory reformulation of labour market relations and the theorising of uneven development at sub-national, regional and local scales again focuses on 'place' and, through a preoccupation with global-local relations, identifies the 'dangers' of the hollowing out of the national. It offers an insight into the nature of the causal processes which underpin local labour market structure and, most importantly, it begins to make explicit the nature of the intersection and construction of those processes within any particular place or geographical context.

This, we would argue, is a major step forward in our understanding. However, this development has not surprisingly exposed potential conflicts in issues of space and place which can not be resolved simply through a redefinition of the local labour market or by and undoubtedly necessary, reconceptualisation of regulation theory to incorporate the sub-national. For the local labour market to retain relevance for the grounding of theory, we would contend that there is a more fundamental relationship to resolve, namely that between existing geographical theories of the activities and operations of the *business enterprise* and those geographical theories relating to the *labour process*, critically as they both operate simultaneously within and between different spatial scales and contexts.

Harvey's (1989b) elaboration of the idea of 'structured coherence' demonstrates the issues of conflicting spatiality that need to be resolved in the local labour market context. 'Structured coherence' is seen as a tendency within an urban economy, reflecting a dominant technology and dominant set of class relations, which arises from the mutual dependency between the daily exchange of labour power and the daily reproduction of labour power caught within the confines of a loosely defined urban labour market. The object of the class alliance is to preserve or enhance achieved models of production and consumption, temporarily for the benefit of both capital and labour though through admittedly asymmetrical power relations, and manifested within a particular location or urban region.

Structured coherence is, in part, an attempt by Harvey to ground abstract Marxism in 'place' by re-emphasising local class relations between capital and labour at the expense of capital-capital relations that occur within a range of spatialities that are much more than just local. We contend that structured coherence in this context creates a limiting and restricted sense of place in which a false urban coalition and unity of purpose is presented, clearly far beyond the 'accidental' alliances it envisages. In Harvey's (1993) words, 'places ... differentiate themselves from other places and become more competitive (and perhaps antagonistic and exclusionary with respect to each other) in order to capture or retain capital investment' (p. 8). Harvey is, of course, explicit in his concern not to present 'the power of place' as if places possess causal powers, and yet he still presents as problematic the relationship between power vested in places and power vested in, for example, large corporations.

> And while there are innumerable signs of decentralization of power to places, there is simultaneously a powerful movement towards a reconcentration of power in multinational corporations [TNCs] and financial institutions. The exercise of this latter power has meant destruction, invasion and restructuring of places on an unprecedented scale. (Harvey, 1993, p. 24)

Our contention is that this distinction between the decentralisation of power to place and the reconcentration of power within TNCs is unnecessary, and somewhat artificial and contrived. The power contained within TNCs is vested in individuals who are, in turn, embedded in 'place'. Within this conception, key actors in terms of power relations operating within a local labour market may equally be influential and a key element within the information and key decision-making structures and networks of corporations, be they truly transnational or only multi-domestic. To conceive of these same individuals' actions as both located intimately within *and* at the same time dislocated from place is unnecessary and ultimately unhelpful to our understanding of both process and place.

Peck's analysis of local labour market processes within a regulation theory framework tends to support this contention and reintepretation. His recent work on Training and Enterprise Councils in the UK (Peck, 1992, 1995) can be interpreted in terms of the conflicting spatialities that arise from the intersection of causal processes within a particular geographical context that create locally uneven development. The Conservative government's TEC initiative involved the handing over of the control of the British training system principally to private sector employers. The Boards of Directors of the TECs are dominated by representatives of British industry who are intended to represent 'the nation's top business leaders; the cream of the local community ... [and] ... should broadly reflect the mix of commerce and industry in their area' (Bennett et al, 1994, p. 9). Here is a contradiction which is best expressed as a conflict of spatialities. On the one hand, the directors are representatives of the elite of British industry and, as such, are linked into 'placeless' corporate and business enterprise networks which are literally 'dis-located' from the specificities of place. On the other hand, these same directors are intimately tied into and are responsible for the development of local training systems and, consequently, the production and direct reproduction of labour power futures for specific local labour markets and local economies (defined crudely and sometimes arbitrarily as groupings of administrative areas). What is apparent, however, in this work on TECs is that while importantly Peck recognises the causal components of labour market structure and emphasises occupational segmentation arising from the intersection of labour demand, supply and state activities, he has also prioritised capital-labour relations and an undefined 'local' spatiality at the expense of capital-capital relations and their multi-layered spatialities.

However, this analysis of TEC directors' potentials for conflict of purpose is illustrative of a wider set of conflicts, in particular, the conflict between transnational corporate power and the economic policies and strategies of nation states which has been well documented over many years. We have, of course, a strong commitment within geography to spatial divisions and uneven development on the basis of the restructuring of capital and its implications for labour. From the perspective described here, however, it may be equally important to understand how that same transnational corporate power and allegiance may be in conflict with local business enterprise, embedded and dependent on 'place', through the intersection of corporate power networks with local economies and markets, and as expressed through the actions and reactions of key actors.

What we argue in the context of local labour markets is that the separation of supply and demand components of the 'local' has led to a position where processes which are central to the construction of 'place' are obscured. To search for a new term for the local which embraces local *and* non-local aspects of the economy and labour market is a minefield to be avoided. The task of unpacking the conflicting topographies of place and space - the conflicting spatialities of capital and labour - is much more than drawing new boundaries which are capable of embracing the 'global' as it impacts upon the 'local'. As Massey (1993) has suggested, in a manner which fits closely with the argument outlined above, and in relation to the conceptualisation of place:

> Definition in this sense does not have to be through simple counterposition to the outside; it can come, in part, precisely through the particularity of linkage *to* that 'outside' which is therefore itself part of what constitutes the place. (p. 67)

Linking the local labour market to the local economy through the intersection of corporate power networks with enterprise embeddedness in place represents one means of reconceptualising place to accommodate the mix of wider and more local social relations which produce place specificity. As Massey (1993) argues, we need a global sense of place which recognises how the apparently 'dis-located' decisions of transnational corporations are vested in individuals who are equally as located in place as the small firm local entrepreneur.

Certainty, stability and spatial fixity

If the essence of 'spatiality', as it is argued here, is its integration into process (as knowledge), rather than as a topography across which events happen or as a series of 'places' as containers of events then, in concrete situations, conflicting spatialities might be reckoned to create a myriad, unpredictable economic-geographic outcomes. That might certainly be the conclusion drawn from the critique of Clark's (1993, 1994) and Schoenberger's (1994) work on strategic management, and it could well be suggested as factor behind the indeterminacy of spatial economic outcomes that is evident in the globalisation literature that highlights global-local interdependencies and tensions. It might even be suggested that 'spatiality as knowledge and information' can only generate a morass of contingency which renders impossible any explanation or real appreciation of change in space economies.

We argue here, however, that this is not the case, and to we suggest that it is possible to identify processes operating at a range of scales within economies and societies that engender and promote *economic stability*, quiescence and certainty within capitalist systems rather than stimulate and accelerate a treadmill of restructuring and change. These processes, we suggest, have been under-valued in the currently dominant political economy approach in economic geography.

The 'new economic geography', with its strong political economy emphasis, tends to be preoccupied with economic transformation and crisis. All too frequently, crisis is seen as the norm and periods of stability and certainty as no more than moments or even as accidents, especially in a world of intensifying globalisation and economic 'disorder'. As Ekinsmyth et al (1995) there is an:

> implicit ideology that underpins much of the 'new economic geography', that stability and certainty are impossible in situations where inequality,

70

domination and subordination exist ... [which] ... derives directly from the Hegelian Marxian dialectical process of development and contradiction. (Ekinsmyth et al, 1995, p.291)

They go on to assert however that:

> Inequality does not necessitate continual change, it just makes truces (negotiated quiescence) [between those exercising power and those resisting] more difficult to achieve. (ibid, p.291)

According to this thesis, which draws on Clegg's (1989) analysis of power, sanctioned inequality is the normal rather than the exceptional state of affairs in the nature of the relationship which binds individuals and enterprises into networks of production and economic activity. It is an argument that runs counter to the notions of embeddedness, trust and reciprocity which underpin current propositions on flexibility, flexible accumulation, new industrial spaces and 'institutional thickness'. It is a set of propositions which recognise tolerated inequality underpinning even medium-term or longer-term (yet still only temporary) stability in a way that is much more than the momentary, accidental tendency of Harvey's (1985) 'structured coherence' because it is needed to enable production and reproduction of both the economy and society - simply to let people live. Indeed, Polanyi (1957, p.251) recognised that stability and unity in societies could be achieved through the integration of actors and agents through both 'redistributional' and 'exchange' relationships involving strong power asymmetries, as well as through 'reciprocal' integration based on equity, trust and loyalty.

Ekinsmyth et al (1995) maintain that just as important as tendencies to crisis in industrial space economies are processes that engender stability and a degree of certainty. They develop their argument in relation to:

- inter-personal and intra-organisational relationships;
- business enterprise structure;
- inter-organisational collaboration and coordination; and
- regulation and governance.

In corporate decision making, it has been suggested that individuals use analogy and metaphor for problem solving which involves recycling experience and resurrecting old problem solving schema - Cyert and March's (1963) 'friction of imperfection' (Sapienza, 1983; Schwenck, 1988). These approaches to strategic decision making can only inhibit change and perpetuate the past. This conservatism in decision making is, in turn, reinforced by the resistance to the exercise of power that is experienced in all organisational hierarchies - a built-in dampener to change (Clegg, 1989).

71

At the level of the business enterprise there is a wide range of processes that create and sustain stability and certainty of which four have been elaborated by Ekinsmyth et al (1995). Firstly, the structure of a business enterprise itself has been recognised as engendering stability, with a firm's 'core technology' being protected from the exigencies of the external economic environment by a buffer of 'boundary spanning structures' to provide certainty and predictability to the 'core' (McDermott and Taylor, 1982). Secondly, as Cyert and March (1963) recognised, 'organisational slack' in an enterprise '... absorbs a substantial share of potential variability in a firm's environment ... and plays a stabilizing and adaptive role' (p. 38). Thirdly, sunk costs and high start-up costs can promote locational and spatial fixity - stabilisation through inertia (Clark, 1994; Baumol et al, 1988). Finally, 'isomorphism', the tendency for business enterprises of a particular era all coming to look the same and to be structured in similar ways, can engender stability (Stinchcombe, 1965). Work regimes in particular eras, mimetic behaviour among firms, fashions in organisational design and even the 'credentialising' of labour can foster a stabilised isomorphism that inhibits change.

At the inter- and supra-organisational scales stability and certainty are fostered by additional processes (Ekinsmyth et al, 1995). First, network relationships, the formation of cartels and the creation of a wide range of strategic alliances are all ploys to avoid uncertainty in economic environments. Second, pressures for the creation of trading blocs are anti-competitive pressures to secure certainty, predictability and stability. Third, real regulation and the institutionalisation of rule fixing to control the emergence of competition and to limit those who are allowed to compete and on what terms are all additional mechanisms used to achieve stability in uncertain economic environments.

What we argue here, therefore, building on the discussion of this section and combining it with notions of conflicting spatialities developed in the earlier sections of the chapter is that processes within economies and societies that promote and foster stability and tolerated inequality also generate *spatial fixity* - a stabilised conflicting spatialities and the production of dynamically stable 'places' which, in very generalised, fuzzy terms might be definable on maps as localities with particular mixes of enterprise structures and labour conditions - industrial districts with unique enterprise formations, capital-labour relations and modes of social regulation.

Conclusion

The starting point for the argument that we have attempted to elaborate in this chapter is that global-local interdependencies are not just an issue of the 'global', 'national' and 'local' as separate spatial spheres of social organisation and action, but are (following Amin, 1997) an issue of the interconnectedness, linkage and hybridisation between agents and places. In Massey's (1993) explicitly geographic terms, a 'place' is constituted by the particularities of its linkages (i.e. the linkages of individuals and collective agencies in that place) to an 'outside' which is more than a simplified territorial caricature. What has been attempted in this chapter is a greater unpacking of the 'spatialities' associated with this interconnectedness and linkage that constitutes places in a globalising world economy. In particular, the role of individual and collective agency in this process has been emphasised - the agency of individuals and the collective agency of business enterprises, organisations and institutions.

Through examples and discussion in the context of strategic management and the decision making of individual managers, the structure, conduct and performance of business enterprises, and the nature and functioning of local labour markets, an attempt has been made to demonstrate the *spatiality of interconnection* is neither a spatial game board across which aspatial actions are played nor a spatial process in its own right. Rather, it has been interpreted as spatiality *within* actions, i.e. within the actions of individual actors and collective agencies.

Central to the argument of the chapter is that these *spatialities within actions* of individuals, individuals within collective agencies and between collective agencies will almost invariably be in a state of tension and *conflict*. The consequence of such tension and conflict is that the geographical patterning and outcomes of global-local interdependencies will, at first sight, create a mass of contingency, randomness, unpredicatability and indeterminacy. However, by recognising the importance of stabilising processes in economies and economic relationships, rather than the crisis tendencies that have been the primary preoccupation of analyses of geographical industrialisation in the past two decades, it is possible to recognise mechanisms within the economies and societies of places that can create 'spatial fixities' - temporarily stable geographical patterns. These patterns might be seen as new spatial divisions of labour in which people and places are defined in terms of their connections with other people and places, and are much more than mere accidents and moments

What we have then is a counter-positioning of tendencies to crisis and tendencies to stability in economies and societies, in much the manner postulated in complexity theory, with the actors and agencies carrying unique spatialities within their actions. Only too clearly, therefore, the linear models of change and spatial patterning that abound in the geographical literature (for example in relation to the new international division of labour and technological change) are difficult to sustain. It may well be through complexity theory that a clearer understanding of global-local interdependencies is to be gained.

Note

1.   Spatiality can be defined in a variety of ways, with the potential to introduce 'sloppiness' into analysis (Gregson and Lowe, 1995). Our definition follows that of Keith and Pyle (1993) and Gregson and Lowe (1995).

References

Amin, A. (1997), 'Placing Globalisation', paper presented to the RGS-IBG Annual Conference, Exeter, 7-9 January.

Amin, A. and Thrift, N. (1994), 'Living in the global', in Amin, A. and Thrift, N. (eds) *Globalization, Institutions, and Regional Development in Europe*, Oxford University Press, Oxford, pp. 1-22.

Bartlett, C.A. and Ghoshal, S. (1989), *Managing Borders: The Transnational Solution*, Harvard Business School Press, Boston MA.

Baumol, W., Panzar, J. and Willig, R. (1988), *Contestable Markets and the Theory of Industrial Structure*, Harcourt Brace Jovanovich, New York.

Bennett, R., Wicks, P. and McCoshan, A. (1994), *Local Empowerment and Business Services: Britain's Experiment with Training and Enterprise Councils*, UCL Press, London.

Buttimer, A. and Seamon, D. (eds) (1980), *The Human Experience of Space and Place*, Croom Helm, London.

Callon, M. (1986),'Some elements of a sociology of translation: domestication of the scallops and the fishermen of St Brieuc Bay', in Law, J. (ed.), *Power, Action and Belief: A New Sociology of Knowledge*, Routledge, London, pp. 196-223.

Channing Adams, P. (1995).'A reconsideration of personal boundaries in time-space', *Annals of the Association of American Geographers*, vol. 85, pp. 267-285.

Christopherson, S. (1993), 'Market rules and territorial outcomes: the case of the United States', *International Journal of Urban and Regional Research*, vol. 17, pp. 272-284.

Clark, G. (1992),'Special issue on "real" regulation', *Environment and Planning A.*, vol. 24, pp. 615-627.

Clark, G. (1993), 'Costs and prices, corporate competitive strategies and regions', *Environment and Planning A*, vol. 25, pp. 5-26.

Clark, G. (1994), 'Strategy and structure: corporate restructuring and the scope and characteristics of sunk costs', *Environment and Planning A*, vol. 26, pp. 9-32.

Clegg, S. (1989), *Frameworks of Power*, Sage, London.

Clegg, S. (1990), *Modern Organizations: Organization Studies in the Postmodern World*, Sage, London.

Coombes, M., Green, A. and Owens, D. (1988), 'Substantial issues in the definition of localities: evidence from sub-group local labour market areas in the West Midlands', *Regional Studies*, vol. 22, pp. 303-318.

Corbridge, S., Thrift, N. and Martin, R. (eds) (1994), *Money, Power and Space*, Basil Blackwell, Oxford.

Crook, S., Pakulski, J. and Waters, M. (1992), *Postmodernization: Change in Advanced Society*, Sage, London.

Cyert, R. and March, J. (1963), *A Behavioral Theory of the Firm*, Prentice Hall, Englewood Cliffs NJ.

Dicken, P. (1992), *Global Shift: The Internationalization of Economic Activity*, Paul Chapman Publishing, London.

Dicken, P. (1994a), 'Global-local tensions: firms and states in the global space economy', *Advances in Strategic Management*, vol.10B, pp. 217-247.

Dicken, P. (1994b), 'The Roepke Lecture in economic geography: global-local tensions: firms and states in the global economy', *Economic Geography*, vol. 70, pp. 101-128.

Dicken, P., Forsgren, M. and Malmberg, A. (1994), 'Local embeddedness and transnational corporations', in Amin, A. and Thrift, N. (eds) *Globalization, Institutions, and Regional Development in Europe*, Oxford University Press, Oxford, pp. 23-45.

Ekinsmyth, C., Hallsworth, A., Leonard, S. and Taylor, M. (1995), 'Stability and instability: the uncertainty of economic geography, *Area* **27**

Florence, P.S. (1948), *Investment, Location and Size of Plant*, Cambridge University Press, Cambridge.

Franko, L.G. (1976), *The European Multinationals*, Harper and Row, London.

Gertler, M. (1992), 'Flexibility revisited: districts, nation states and the forces of production', *Transactions of the Institute of British Geographers, New Series*, vol. 17, pp. 259-278.

Giddens, A. (1996), 'Affluence, poverty and the idea of a post-scarcity society', *Development and Change*, vol. 27, pp. 365-377.

Goodman, E., Bamford, J. and Sayner, P. (eds) (1989), *Small Firms and Industrial Districts in Italy*, Routledge, London.

Gould, P. and White, R. (1974) *Mental Maps*, Penguin, Harmondsworth.

Gregson, S. and Lowe, M. (1995), ' "Home"-making: on the spatiality of daily social reproduction in contemporary middle class Britain', *Transactions of the Institute of British Geographers, New Series*, vol. 20, pp. 224-235.

Handy, C. (1994), *The Empty Raincoat: Making Sense of the Future*, Hutchinson, London.

Harvey, D. (1985), *The Urbanization of Capital*, Basil Blackwell, Oxford.

Harvey, D. (1989a), *The Urban Experience*, Basil Blackwell, Oxford.

Harvey, D. (1989b), *The Condition of Postmodernity*, Basil Blackwell, Oxford.

Harvey, D. (1993), 'From space to place and back again: reflections on the condition of postmodernity', in Bird, J. et al (eds) *Mapping the Futures: Local Cultures, Global Change*, Routledge, London, pp 3-29.

Held, D. (1995) *Democracy and the Global Order*, Polity, Cambridge.

Hirst, P. and Thompson, G. (1992), 'The problem of globalisation: international economic relations, national economic management, and the formation of trading blocs', *Economy and Society*, vol. 21, no. 4, pp 357-396.

Hirst, P. and Thompson, G. (1996), 'Globalisation; ten frequently asked questions and some surprising answers, *Soundings*, vol. 4, autumn, pp. 47-66.

Jessop, B. (1990), 'Regulation theories in retrospect and prospect', *Economy and Society*, vol. 19. pp. 153-216.

Keith, M. and Pyle, S. (eds) (1993), *Place and the Politics of identity*, Routledge, London.

Lazerson, M. (1993), 'Factory or putting out? Knitting networks in Modena', in Grabher, G. (ed.) *The Embedded Firm: In the Socioeconomics of Industrial Networks*, Routledge, London, pp. 203-226.

Martin, R. (1994), 'Stateless monies, global financial integration and national economic autonomy: the end of geography?', in Corbridge, S., Thrift, N. and Martin, R. (eds) *Money, Power and Space*, Blackwell, Oxford, pp. 253-278.

Massey, D. (1984), *Spatial Divisions of Labour: Social Structures and the Geography of Production*, Macmillan, Basingstoke.

Massey, D. (1993), 'Power-geometry and a progressive sense of place', in Bird, J. et al (eds) *Mapping the Futures: Local Cultures, Global Change*, Routledge, London, pp. 59-69.

McDermott, P. and Taylor, M. (1982), *Industrial Organisation and Location*, Cambridge University Press, Cambridge.

McGrath-Champ, S. (1995), 'The Nature and Role of Strategy in Industrial Restructuring', paper presented to the IGU Commission on the Organisation of Industrial Space Conference, "Interdependent and Uneven Development: Global-Local Perspectives", Seoul, 7-11 August.

Ohmae, K. (1990), *The Borderless World: Power and Strategy in the Interlinked Economy*, Stanford University Press, Stanford).

Peck, J. (1989), 'Reconceptualizing the local labour market: space, segmentation and the state', *Progress in Human Geography*, vol. 13, pp. 42-61.

Peck, J. (1992), 'TECs and the local politics of training', *Political Geography*, vol. 11, pp. 335-354.

Peck, J. (1995), 'Geographies of governance: TECs and the remaking of 'community interests', paper presented to the Institute of British Geographers annual conference, University of Northumbria, Newcastle upon Tyne, 3-6 January.

Peck, J. (1996), *Work-place: The Social Regulation of Labor Markets*, The Guilford Press, New York.

Pfeffer, J. and Salancik, G. (1978), *The External Control of Organizations: A Resource Dependence Perspective*, Harper and Row, London.

Polanyi, K. (1957), 'The economy as an instituted process', in Polanyi, K., Arensberg, C. and Pearson, H. *Trade and Markets in Early Empires* Free Press, Glencoe ILL, pp. 243-270.

Porter, M.E. (ed.) (1986), *Competition in Global Industries*, Harvard Business School Press, Boston.

Powell, W.W. (1990), 'Neither markets nor hierarchies: network forms of organization', *Research in Organizational Behaviour*, vol. 12, pp. 295-336.

Prohansky, H. (1978), 'The city and self-identity', *Environment and Behaviour* vol. 10, pp. 147-169.

Redclift, M. and Benton, T. (eds) (1994), *Social Theory and the Global Environment*, Routledge, London.

Reich, R. (1991), *The Work of Nations: Preparing Ourselves for 21st Century Capitalism*, Vintage Books, New York.

Sabel, C.F. (1989), 'Flexible specialization and the re-emergence of regional economies', in Hirst, P. and Zeitlin, J. (eds) *Reversing Industrial Decline?*, Berg, Oxford, pp. 17-71.

Sapienza, A. (1983) 'A cognitive perspective on strategic formulation, paper presented to the Academy of Management Meetings, Dallas, Texas.

Seamon, D. (1979), *A Geography of the Lifeworld*, St Martin's Press, New York.

Schoemaker, P. (1993), 'Strategic decisions in organizations: rational and behavioural views', *Journal of Management Studies*, vol. 30, pp. 107-129.

Schoenberger, E. (1994), 'Corporate strategy and corporate strategists: power, identity and knowledge within the firm', *Environment and Planning A*, vol. 26, pp. 435-451.

Schwenck, C. (1988), 'The cognitive perspective on strategic decision making', *Journal of Management Studies*, vol. 25, pp. 41-55.

Stinchcombe, A. (1965), 'Social structure and organizations', in March, J.G. (ed.), *Handbook on Organizations*, Rand McNally, Chicago.

Storper, M. and Harrison, B. (1991), 'Flexibility, hierarchy and regional development: the changing structure of industrial production systems and their forms of governance in the 1990s', *Research Policy*, vol. 20, pp. 407-422.

Swyngedouw, E. (1992), 'The mammon quest. 'Glocalisation', interspatial competition and the monetary order: the construction of new scales', in Dunford, M. and Kafkalis, G. (eds), *Cities and Regions in the New Europe*, Belhaven, London, pp. 39-67.

Taylor, M. (1987), 'Technological change and the business enterprise', in Brotchie, J. Hall, P. and Newton, P. (eds) *The Spatial Impact of Technological Change*, Croom Helm, London, pp. 208-228.

Taylor, M. (1994), 'The business enterprise, power and patterns of geographical industrialisation', in Conti, S., Malecki, E. and Oinas, P. (eds) *The Industrial Enterprise and Its Environment: Spatial Perspectives*, Avebury, Aldershot, pp. 99-122.

Thrift, N. (1986) 'The geography of international economic disorder', in Johnston, R.J. and Taylor, P.J. (eds) *A World in Crisis: Geographical Perspectives*, Blackwell, Oxford, pp. 12-67.

Tuan, Y-F. (1975), 'Place: and experiential perspective', *Geographical Review*, vol. 65, pp. 151-165.

Tuan, Y-F. (1977), *Space and Place: The Perspective of Experience*, Edward Arnold, London.

Tuan, Y-F. (1982), *Segmented Worlds and Self: Group Life and Individual Consciousness*, University of Minnesota Press, Minneapolis.

Whittington, P. (1988), 'A realist approach to strategic decision-making', *Journal of Management Studies*, vol. 25, pp. 39-54.

4 Commodity system governance by quality management: a New Zealand discourse

Richard Le Heron, Ian Cooper, David Hayward and Martin Perry

Introduction

This chapter explores emerging ideas about regulatory arrangements connected with global eco-commodity systems with particular reference to the development of quality management structures in the horticultural sectors of the Hawkes Bay, New Zealand. It begins by exploring an apparent convergence in interest and theoretical ideas relating to global commodity systems in several relatively independent literatures. From the review a range of overlapping ideas in the global restructuring and networks literatures are identified as a basis for a revised conceptualisation of the production-commodity system idea. However, in the contributions of economic geographers, sociologists, anthropologists and economists, the actual *geography* of global commodity systems has been largely overlooked. This missing dimension has major implications for the treatment of regulatory questions and the interpretation of growth and globalisation associated with commodity system development. The chapter sketches a framework for situating and interpreting the intersecting processes of growth and regulation. It then outlines the rise of total quality management in New Zealand's horticultural complex as a backdrop to the investigation of networking and networks connected with quality management issues in five horticultural sectors in the Hawkes Bay region. Quality management is viewed as a composite strategy adopted by a range of agents to deal with changing consumer preferences and emerging regulatory frameworks. The findings reveal how quality management developments embody significant elements of governance with a range of impacts on and implications for the evolution, coherence and territorially embedded social relations of global commodity systems.

Converging theoretical ideas: an overview

Four trajectories on the idea of production-to-consumption sequences under capitalism can be identified in the development of theory. These trajectories have developed in industrial geography, business management, the political economy of agriculture and world systems/political geography. They are shown schematically in Figure 4.1. Leaving aside the finer genealogy of each trajectory and the problems of putting labels on research which may have less coherence than the diagram would imply, some common ground can be found in these broad intersecting areas of theoretical enquiry. In particular, each body of literature addresses in varying ways the mutual interdependence of organising agents and patterns of organisation, a focus which is central to exploring the politicised processes of organisation and stabilisation associated with production-to-consumption sequences.

Before exploring these areas of theory two comments about the vocabulary of this section need to be made. First, the term 'global commodity system' is left undefined until the last area of theory has been discussed. This is because the term has particular origins, and can be strengthened from the combination of insight from several literatures. Second, discussion of the geographical dimensions of global commodity systems is postponed until the next section. This reflects the absence of readily available concepts in all the areas of research discussed here that are needed to explore the geography of particular systems.

In the contributions of industrial geographers, (for example, Dicken, 1993, 1994; Walker, 1988) the notion of production and circulation systems as distinguishing features of capitalist industrialisation have been stressed. This conceptual framework has guided analyses of industrial growth, which, while generally set in the context of institutional structures, have tended to under-theorise the role of the state and regulation. In the main there has been little explicit treatment of either the wider regulation or the wider geography of production-consumption systems. This can be accounted for in part by the national or comparative-national orientation of research. Most research in the 1990s has concentrated on the embeddedness of production developments, especially networking, in industrial regions (Dicken, et al, 1994; Storper, 1992). Recent research has highlighted the consumption end of the production system on the grounds that this is where economic power is presently concentrating. Analyses in this vein are for the most part preliminary. They verge on ignoring wider dimensions of transformation that characterise contemporary commodity systems and may be unwittingly bound by the national regulatory setting in which restructuring is occurring (Parsons, 1995). A tentative restatement of the production system concept as a production-consumption system involving both production and consumption sites is found in Le Heron and Park (1995). Further, Fagan and Le Heron (1994) provide a general framework for analysing the circulation of capital and its regulation in and through territory, which has applicability to commodity system analysis.

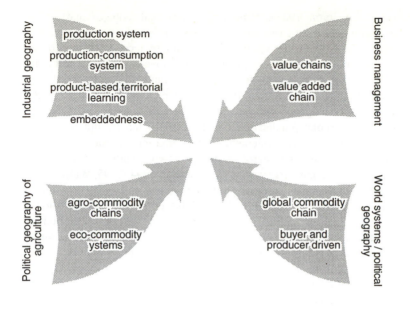

Figure 4.1   Converging theoretical positions

Parallelling the contributions of industrial geographers are the extensive writings on value chains (Porter, 1990). Indeed Dicken and others acknowledge ideas from this body of knowledge. This school of inquiry has emphasised the movement through the production-consumption sequence of value and the sites where most value can be or is added (Crocombe, Enright and Porter, 1991). This emphasis has encouraged business researchers with an interest in networking to consider the wider power relations in which networking is proceeding (Axelsson and Easton, 1992). Researchers in this field have been receptive to how adjustments can impact (or should impact, in the proactive language of quality management and assurance) throughout the value chain. Nonetheless, discussion does not extend to consideration of the social dimensions of value chain restructuring or transcend the general interest in international competitiveness as distinct from looking at the nature of global connections implied by deepening incorporation into the global economy.

Within the political economy of agriculture writings the appearance and transformation of agro-commodity systems has attracted attention (Friedland, 1984; Goodman, Sorj and Wilkinson, 1987; Le Heron, 1993; McMichael, 1994). These writings have gone furthest in developing a framework that gives theoretically balanced treatment to accumulation and regulatory issues relating to agro-commodity systems. The importance of different governance arrangements is clearly recognised, and the historical specificity of agro-commodity system evolution has been documented (Le Heron, 1988). However, the national scope of analysis (for very practical reasons) in most cases has meant that while links to the global economy have been examined the wider interactions implied by incorporation into the world economic system have largely been put aside. To move beyond (but still regain) the nation state, either with respect to growth dynamics or regulatory structures, a conceptualisation of agro-commodity systems which is global at the outset, yet confers flexibility to probe national and local articulations, is essential. The beginnings of a reconceptualisation can be constructed by drawing on several prominent ideas in the world systems/political geography literature.

Within world systems research a strand of work addresses the long period evolution of global commodity systems (McMichael, 1994, 1995; Wallerstein, 1991) against the backdrop of core-periphery relationships amongst nation states. According to Gereffi et al (1994, p. 13), 'global commodity chains allows ... [a] focus on the creation and distribution of global wealth as embodied in a multidimensional, multistage sequence of activities, rather than as the outcome of industrialization alone'. Ignoring for the moment the obvious omission of 'multilocational' from the sequence idea in the quote, the thrust of the conceptualisation is overtly global. As a partial architecture for investigating the global circulation and regulation of capital, the global commodity chain seems to offer a promising start. Gereffi (1994) attempts to show a shift in the power relations inside global commodity chains, distinguishing the structure of chains according to whether the primary guiding influences come from buyer pressures or from producer action. His discourse, however, is only coarsely spatial, missing the possibility of constructing differently situated perspectives on developments characterising the commodity systems (Le Heron and Roche, 1995).

Approaching a geography of global commodity systems

Importantly, each of the four theoretical trajectories outlined above recognises, to varying degrees, the resolution of the social relations of production and consumption relations in territory. This idea, while still largely implicit in much research, is equally applicable to either production or consumption circuits, or any intermediary steps in the sphere of circulation spanning those spheres of activity. This interpretation, formulated in a neo-productionist mould in Fagan and Le Heron (1994), acknowledges the mediating influences of the state but fails to generalise the proposition into the idea of mediating regulatory structures (e.g. government, non-government, intra-firm) operating at a range of geographic scales. In order to tease out the regulatory aspects of global commodity chains, further conceptualisation is required. In particular, the systemic features of commodity chains, as constituted at sites of production, circulation and consumption need outlining. This can be accomplished by utilising the concepts of globalisation, localisation, linkages and networks. Figure 4.2 conceptualises the key relationships of global commodity systems using these concepts. Globalisation is depicted as processes operating at the

84

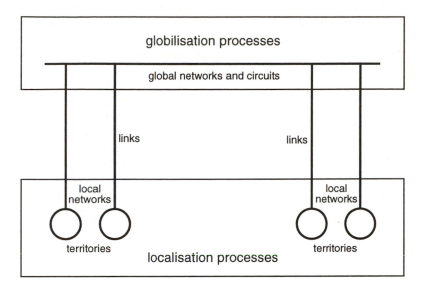

Figure 4.2 Conceptualising the geography of global commodity systems

transnational scale and constituted by and constitutive of production and consumption activities in different territories. Production and consumption sites are linked to and through the processes of globalisation. Localisation in contrast summarises processes manifest at micro scales. The mediated two-way flows in the circulation sphere, bridging and interpenetrating global and local processes, generate differentiated outcomes and experiences at every point in the global commodity chain. Moreover, the changing nature of links, to and from territories gives geographical and historical distinctiveness to the expression of globalisation and localisation in specific areas. Activities relating to both production and consumption are embedded in local networks and simultaneously connected through global networks and circuits. The switch from producer-driven to buyer-driven global commodity systems, suggested by Gereffi (1994), is accommodated by the conceptualisation. Indeed, the conceptualisation is central to a fuller understanding of capitalist commodification and the systemic development of capital circulation. Thus, conceptually, global commodity systems are inherently geographical in their structuring, but analyses have generated few details of the geography of this structuring.

85

Global commodity systems, global restructuring and networking

Global commodity systems are relatively stable transnational arrangements for the circulation of capital, constructed from the complex and changing power relations of the organisations of commerce and government through which accumulation occurs. The historical persistence of arrangements has been facilitated, and on occasion hindered, by both public and private regulatory frameworks, of varying nature, complexity and geographical reach. At the global level and in different national contexts, capitalism has featured historical shifts between market regulation and political regulation of economic processes (Arrighi, 1994; Beck, Gregory and Lash, 1994). These polar forms of regulation appear to involve a drift beyond industrial concentration toward concentration of control and coordination of capital circulation. In the former case, centralisation increasingly redistributes structural power to finance capital via market processes, while in the latter, centralisation is accomplished and exercised through political processes (Leyshon, 1992). Thus, global commodity systems must be seen as nested in wider geopolitics, national political structures, as well as outgrowths of enterprise decision-making, all incorporating elements of governance. Pressures derived from systemic developments in a range of contexts find expression in public sector regulatory adjustments, such as the restructuring of the state apparatus of many nations, the formation of multinational regulatory blocks and the emergence of the enterprise-oriented local state. Private sector adjustments are equally comprehensive, encompassing restructuring of corporate ownership and control in commodity chains.

Over capitalism's history, some particularly enduring and important transnational axes of investment are discernible. These international regulatory regimes have been variously labelled as resource regimes or food regimes, for example (Friedmann and McMichael, 1989). Regimes are composites of public and private regulatory arrangements, aimed at containing the contradictions of particular global commodity systems and territories in which concentrated production circulatory activity and consumption is found. They are constructed in and from particular territorial contexts but necessarily transcend those contexts in order to secure satis-factory long term arrangements for the circulation of particular values (Long, 1995). Such arrangements can be regarded as an integral part of global commodity systems.

Two transnational governance solutions broadly underpinned the stability, growth and increasing internationalisation of global commodity systems in the post-1945 long boom period. These were transnational corporations (TNCs) and their equivalent, producer-marketing boards (PMBs), which secured, in particular geographically-grounded articulations of production, differing degrees of vertical integration from production through to consumption. The TNC and PMB frameworks represent hierarchical solutions to governance, in the first case under the umbrella of a single firm, and in the second case as a politically contrived organisational form that confers similar capacities to coordinate segments of each commodity chain. In the end, the key distinction between these organisational solutions is the nature of distributive arrangements - the PMB model offering prospects often unachievable, of democratic resolution of income and profit distribution. Shifts away from more centralised coordination, which is the direction implied by the integrative tendency behind and evident in contemporary global commodity systems, is creating structural space for new governance solutions. The post 1945 order is being rapidly transformed by heightened international competition (Friedmann, 1991, 1993; McMichael, 1994).

A widely documented surge in networking, at all geographic scales, can be interpreted as part of the search for new institutional solutions to resecure competitiveness and stability amidst the turbulence of global restructuring. This is not to suggest that networking per se is either 'the' solution of a 'stable' solution. Rather, networking can be likened to a translating activity, a tangible expression of efforts to negotiate new arrangements in socio-political, cultural-economic contexts which are highly turbulent. In our interpretation, the emphasis must be placed on the active form, *networking,* as this connotes intentionality, whereas *networks* perhaps unfairly suggests static arrangements.

Networking and networks have always been found in the commercial world. Arrighi (1994), for instance, discusses networking responses in the circulation sphere as foreshadowing and accompanying major transformations in capitalism. The late twentieth century is a period of unprecedented levels of networking and network formation. Intensified networking can be accounted for in large measure as a general response to the pressures of restructuring in the world's global commodity systems. Most business networking is ultimately anchored in value creating or preserving activities, blurring the distinction between transactions, contracts and networks. In the 1990s, the politics and power relations of global commodity chain governance seems to be gradually shifting from commodity system coherence, based on acceptance of the rules of transfer laid down by particular strategic actors (PMBs or TNCs), to commodity system coherency derived from the adoption and implementation of rules inherent in systems of international standards. Of course the definition of the new standards, the actual specifications chosen, the procedures for enforcement and sanction, the terms under which conformance might be allowed, the documentation required and so forth, remain set in the international and national politics of market access and in the territorially-based power relations of particular portions of each commodity system. With the directive powers of key organisations being either severely constrained (the PMBs) or exercised in different ways (the TNCs), networking and networks have become important social arrangements for bridging problematised steps in commodity chains and maintaining commodity system stability.

New Zealand's restructuring and networking

In the 1980s New Zealand responded to international economic pressures by embarking on a phase of state-induced restructuring. Re-regulation in New Zealand has proceeded on two fronts, those of the economy and those of the state (Britton, Le

Heron and Pawson, 1992; Le Heron and Pawson, 1996). A major consequence of this has been growing criticism and failure of prevailing models of commodity system governance. This has taken two main forms. First, direct challenges have been made to the PMB model. In the cases of apple growing and dairying the challenge has had little success (Blunden, 1996). The PMB model remains the overwhelming choice of farmers in the face of some orchestrated opposition from corporate interests at the farm, processor and marketing levels. Second, foreign capital, a major integrating force for decades, has withdrawn from some sectors, the most notable being the meat industry. In response to this development, New Zealand meat companies have joined together to formulate a Meat Industry Plan (Lynch, 1996). Thus, in New Zealand, internal adjustments add to external pressures to restructure commodity systems to conform to emerging governance frameworks centred on qualitative measures. The high level of interest in networking is an integral aspect of restructuring responses. Continuing uncertainties over governance alternatives in New Zealand probably means that the networking era in New Zealand may be a lengthy one as competing interests seek to assert different rules on the coordination and control over aspects of different value chains. Networking behaviour is, moreover, likely to change in the face of continuing experimentation. Indeed Sayer (1989), echoing Lipietz (1986), argues that organisational forms are discoveries worked out in different contexts. By this view networking and networks amount to transformative processes, both in and for global commodity systems and regions.

Impetus for planned networking in New Zealand horticulture

When viewed from a commodity chain perspective, significant pressures for regulatory responses face New Zealand's horticultural sectors (Le Heron and Roche, 1996). These pressures emanate from the buyer environment. The apple sector is adjusting to due diligence legislation in the UK, phytosanitary and pesticide protocols in the US, and general international concern about spray residues. Grape growers are meeting tighter standards so that grape varietal features can be enhanced and the appellation system has implications for the monitoring of production by area. Squash producers conform to strict Japanese buyer requirements and asparagus suppliers to Japan are sensitive to buyer specifications. Tomato growers, while producing for processing, are increasingly moving towards low pesticide farming (Cooper et al, 1995).

Since the mid 1980s a new approach to marketing, emphasising TQM has been actively promoted in New Zealand (Perry, 1995; Perry and Goldfinch, 1996; Perry et al, 1995). In horticulture, advocacy of TQM has come from the Food and Beverage Exporters Council (FBEC). The FBEC was established in 1985 as a joint promotional body with the government's Trade Development Board (Tradenz). This group has been active in promoting export-oriented marketing and quality management programmes in the food processing sector. Within the FBEC there is a strong consensus that company growth depends on strong brand identification and total quality through the value chain. One component of the FBEC strategy is a quality advancement programme launched in 1992 in partnership with Tradenz. While the FBEC initiative is primarily targeted at food processors, several participants are directly linked to the Hawkes Bay horticultural sector, including the Apple and Pear Marketing Board (APMB) and J Wattie Industries. The significance of the TQM initiative is that a range of ideas, perceived to be critical to international competitiveness, are relatively well known in the Hawkes Bay.

In important respects, TQM is a transformative philosophy which rests upon network building within and between organisations. TQM was developed in the context of Japanese manufacturing, where it served three main purposes:

- diffusing responsibility for assurance across all members of an organisation, instead of it being the responsibility of a separate group of quality control personnel;
- supporting the reduction of inventory and the implementation of just-in-time management; and
- encouraging an environment of worker-management trust, partly as a way to obtain workforce involvement in continuous incremental innovation.

TQM methods do not identify a single organisational strategy. Rather they provide a package of measures that are employed with varying degrees of emphasis, depending on context.

Transferring TQM principles to all sectors and levels of the horticultural complex is no easy task. Table 4.1 shows seven TQM principles and attempts a summary of how the principles might be applied in horticulture and the implications for networking that might derive from the adoption of such principles. The table points to relationships in the commodity system that are likely to be the site of transformative networking should TQM be actively pursued. The table is a valuable benchmark for assessing the level of development and significance of networking and networks in Hawkes Bay horticulture.

Horticultural sectors and their evolution

Five horticultural activities are important in Hawkes Bay, the growing of apples, grapes, asparagus, squash and tomatoes, with apples and grapes dominating the regional scene. Table 4.2 shows the number of growers, packhouses and processors currently operating in the region. Table 4.3 summarises recent production and planting trends and the proportion of New Zealand sectoral output and area in the sectors investigated. The notable trend is the steady rise and prominence of the apple sector. This sector is the lead sector in the horticultural complex on a range of criteria - plantings, output increases, number of growers, range of products, research infrastructure, capital investment in infrastructure. The area under vines has risen sharply in the past two seasons. The other sectors have had fluctuating fortunes.

Table 4.1
**Total quality management principles and their application to the
horticulture sector**

TQM Principle	Application to horticultural sector	Potential implications for networking
1 Quality over price	Existing marketing	End-buyer
2 Internal customers	Steps in chain	Value - added - retained
3 Meeting specifications	R & D and grower product	Field days, technology knowledge transfer, labour training
4 Quality issues have wide origins	Limited internal workforce	Wider communications
5 Senior management drive acceptance of changed responsibilities	Diffusion across separate organisations	Looking beyond seasonal links
6 Improvement depends upon culture change	Relationships between steps	Grower associations
7 Quality depends on all	Involve previously isolated participants	Development of a permanent horticultural workforce

Source: Extended from Hill (1991)

Table 4.2
Selected Hawkes Bay horticulture sectors, 1995

	Apples	Grapes	Squash	Asparagus	Tomato
Growers	800 +	≈ 200	19	140	8
Packhouses/ Processors	≈ 120/≈5	20 +	≈ 5	5	2
Sales Profile	Europe + US, Asia	NZ (UK+)	Japan	Japan	Local Processors

The 1994-1995 structural situation in the Hawkes Bay sectors is shown in Table 4.4. Each of the five horticultural chains examined - apples, asparagus, grapes, tomatoes and squash - is associated with distinctive grower risks, barriers to entry, availability of product knowledge, export and processing options, grower organisation, marketing structures and integration to the end-buyers. These contextual aspects have a major bearing on how external pressures are perceived and the willingness, capacity and interest of different stakeholders at sites in each commodity chain to collaborate via networking to resolve issues which are seen as threatening chain integrity or particular aspects of production.

Dimensions of networking: governance through quality management?

A two-stage survey of 30 organisations and 42 growers and packhouses in 1994 and 1995 in the horticultural complex of the Hawkes Bay established the contextual aspects and changing nature of networking in the five horticultural chains. The rise of quality management in the sectors was specifically explored with reference to three inter-related dimensions of networking, those of *market development, quality assurance and training and skills.* These dimensions represent different facets of commodity chain reconstruction occasioned by either buyer-driven adjustments to consumer preference shifts or institutional and political demands to meet new international standards. The underlying rationale is thus that the fundamental shift in commodity system power relations is a material influence upon stakeholders. Whether this shift is manifest in the eco-commodity systems of the Hawkes Bay can only be established from empirical investigation.

Table 4.3
The Hawkes Bay horticultural scene

		1981	1986	1990	1992	1993	1994	1995
Apples	Export (10^6 ct)	2.7	4.4	5.8	5.9	6.3	4.1	7.2
	% NZ	51	50	53	53	50	36	na
Grape	(000s t)	-	-	20.3	14.2	8.4	15.1	20.6
	% NZ			29	26	20	28	28
Asparagus	(Total t)	-	-	1660	1710	1430	1890	-
	(Exp t)		83	220	270	210	350	
	% NZ Total			19	22	19	23	
	% NZ Exps		6	14	18	16	22	
Squash	NZ Exps (000s t)	6.5	37	63	74	70	94	69
Tomato	(000s t)	12	14	40	41	25	39	37

Source: Business Networks and Exporting project

Table 4.4
Overview of selected Hawkes Bay horticulture sectors

	Apples	Grapes	Squash	Asparagus	Tomato
Barriers	cost-land	cost-land contracts	rotation, risks	growing risks	contracts
Product Control	High	High	Low	Medium	Medium
Process/Export Options	Medium	Medium	Low	High	Low
Grower Organisations	Mkt Board Allegiance	Industry Association	Divided Allegiances	National and Local Assoc.	Bilateral Grower Reps
Marketing	Mkt Board	Contracts	Annual procurement	Long-term links	Annual contract
Integration	Direct through Marketer	None	None	Direct through Marketer	None

Source: Business

93

Summaries of networking activity on the three dimensions listed above are presented in Table 4.5. Entries in the table deal only with networking which has sprung up during the past five years in response to the growing interest in quality management. Some of the networking builds on existing organisational links, but the entries record new activity amongst groups of growers or grower organisations that is a departure from past practice because of the connection to quality management pressures.

A comparative review of sectors in terms of the level of development of TQM initiatives (broadly interpreted) is shown in Table 4.6. In each sector efforts are clearly being made to incorporate quality management principles into growing, harvesting and post-harvesting practice. Networking has been important in disseminating quality management ideas and practices. The actual pattern of networking and the makeup, duration, procedures and effectiveness of networks brought into existence is very specific to each sector. Thus, the evidence suggests that recognition of quality management as a necessary aspect of horticulture is closely associated with networking. The surveys, however, were conducted at a time when quality management was still in its infancy in each chain so performance of growers and packhouses could not be directly related to involvement in networking. Despite the high incidence of quality-related networking in each chain, a potentially significant gap in networking relates to the horticulture work force. This element forms the seventh principle in Table 1 and is an indicator of how embedded networking is in the region, as distinct from specific commodity system cultures. The underdevelopment of networking on this dimension probably reflects a partial understanding of TQM principles and a failure to appreciate the whole system transformation that is central to the TQM movement. Major tensions about labour were identified in the survey especially at the grower level in each sector. The results show at best a moderate recognition of what is entailed in fostering a skilled regional labour force in horticulture.

Conclusions

The survey revealed that in the Hawkes Bay horticultural complex quality issues and questions are to the fore in each sector. The degree to which different horticultural practices have been instigated is nonetheless highly varied. The dominant apple sector has been forced to adjust to a series of international demands, although the Apple and Pear Marketing Board (APMB) has attempted over the 1994-95 season to turn international regulatory requirements into a force for significant reorganisation of grower and packhouse practices. The stimuli of due diligence and hygiene regulations from the European market and insect and pesticide concerns in the North American sector have been used by the APMB as a basis for developing the US Code of Practice, as a New Zealand inspired 'international standard' for the global apple system. This is not surprising given New Zealand's current capacity to

	Network	Influences (constraining / enabling)
Market Development		
Apple	APMB	
Grape	Wine Guild-Wineries-TRADENZ	National Industry Growth Plan
Squash	Squash Council	Japanese distribution chain
Asparagus	HBAGA	flexibility between export & local market
Tomato	Corporates	Local near-monopoly
Cross-sector	Food & Bev Exp Council-TRADENZ	
Quality Assurance & TQM		
Apple	Overseas customers-APMB packhouses-growers GroSafe	
Grape	Wineries-growers	GroSafe mandated by winery
Squash	Importers-SQ Council-Packhouses-growers Sq Council/BSEA	Compulsory
Asparagus		?
Tomato	Processor-growers	
Cross-sector	Food & Bev Exp. Council-TRADENZ	
Training & Skills		
Apple	GroSafe Spray diaries	Mandated by APMB "
Grape	HBGGA Wineries-growers	
Squash		
Asparagus	HBAGA	
Tomato	Grower groups GroSafe	(Voluntary)
Cross-sector	HB Polytechnic	

Table 4.6
Developments in TQM in Hawkes Bay horticulture

TQM Principles	Apples	Grapes	Squash	Asparagus	Tomatoes
1	high	high	high	fresh-high processed-med	high
2	high	medium	low	high	high
3	high	medium	low	medium	low
4	high	high	medium	high	high
5	high	high	medium	high	medium
6	medium	high	medium	high	low
7	medium	low	low	medium	high

obtain far higher point of sale returns than for instance Chile. Other sectors have examples of quality-consciousness which also illustrate the appearance of new dynamics within the commodity chains.

Networking has developed to facilitate quality management initiatives. Such networking is a mixture of efforts aimed at satisfying externally imposed demands and collaboration to develop distinctive product-process systems as a basis of competitiveness. Aspects of TQM are being actively explored by some organisations, in several sectors, but there is no strong evidence of wholesale sector-by-sector adoption of the package of TQM principles identified. Nevertheless, the extent of response to quality related pressures (in whatever form they are felt) is sufficient to give some sense of how emerging sensitivity to quality is beginning to dictate system-wide transformation. Perhaps the most critical conclusion is that quality management is clearly an ingredient in any new relationship of the Hawkes Bay sectors to the global fresh fruit and vegetable complex. The continued participation by any of the sectors in this complex is always problematic (since the interviews a processor has gone into receivership). The links of local systems into global processes is broadly reconstructed from season to season. But the re-establishment of a link, with attendant two-way global-local interactions can be jeopardised at any time, as the Hawkes Bay apple growers found in the 1993-94 season, when insect counts in apples exported to the US almost exceeded USDA entry limits. Such events can galvanise local support in particular sectors and be a warning to growers in other sectors of the fragility of their commodity system arrangements.

However, unless sector-specific quality management practices are widely and rapidly adopted the possibility of translating structurally-driven pressures into a new basis for international or regional competitiveness remains slim. Indeed, even the suggestion that the sectors make up a *complex* may be too generous an interpretation of the Hawkes Bay scene. While a number of sectors co-exist in the Hawkes Bay, there is little evidence to suggest that cross-sectoral networking is extensive or being widely sought. Where such networking is found it is prototype in nature and yet to stand the test of wider application. Moreover, the general vulnerability of each sector to system shocks, such as price collapses, regulatory rulings or natural hazards, further constrains how far stakeholders are able to coordinate developments through networking. Thus, despite a considerable incidence in networking activity relating to quality management in the horticultural sectors of the Hawkes Bay, no case can be made that the regional economy represents a 'network economy' or an 'intelligent region'.

Acknowledgements

We gratefully acknowledge the support of the Foundation of Research Science and Technology (Contract CO 7404) and the willing assistance given to us by the many people we contacted and interviewed in the Hawkes Bay. Further details of survey findings and networking can be found in Cooper, et al (1995).

References

Arrighi, G. (1994) *The long twentieth century,* Verso, New York.

Axelsson, B. and Easton, G. (eds) (1992), *Industrial Network: A New View of Reality;* Routledge, London.

Beck, G. D. and Lash, S. (1994), *Reflexive modernization,* Stanford University Press,Stanford.

Blunden, G. (1996), 'The debate over the powers of the producer marketing boards', in Le Heron, R. and Pawson, E. (eds), *Changing Places. New Zealand in the Nineties,* Longman Paul, Auckland, in press.

Britton, S., Le Heron, R. and Pawson, E. (eds), (1992), *Changing Places in New Zealand. A Geography of Restructuring.* New Zealand Geographical Society, Christchurch.

Cooper, I., Perry, M., Le Heron, R. and Hayward, D. (1995), Business Networking and Exporting. A Profile of Networking and Networks in the Apple, Asparagus and Grape, Tomato and Squash Sectors of the Hawkes Bay Horticultural Complex, Occasional Paper No. 31, Department of Geography, University of Auckland, Auckland.

Crocombe, G. T., Enright, M. J. and Porter, M. (1991), *Upgrading New Zealand's Competitive Advantage,* Oxford University Press, Auckland.

Dicken, P. (1993), 'The changing organisation of the global economy', in Johnston, R. J. (ed) *The Challenge for Geography,* Blackwell, Cambridge, pp. 31-53

Dicken, P. (1994), 'The Roepke Lecture in Economic Geography. Global-Local tensions: firms and states in global space economy', *Economic Geography,* vol. 70, no. 2, pp. 101-128.

Dicken, P., Forsgren, M. and Malmberg, A. (1994), 'The local embeddedness of transnational corporations' in Amin, A. and Thrift, N. (eds), *Globalisation, Localisation: Possibilities for local prosperity,* Oxford University Press, Oxford.

Fagan, R. and Le Heron, R. (1994), 'Reinterpreting the geography of accumulation: the global shift and local restructuring', *Environment and Planning D, Society and Space,* vol. 16, no. 3, pp. 265-285.

Friedland, W. (1984), 'Commodity systems analysis: an approach to the sociology of agriculture', in Schwartzweller, H. K. (ed.), *Research in Rural Sociology and Development,* JAI Press, Greenwich, CT, vol. 1, pp. 221-236.

Friedmann, H. (1991), 'Changes in the international division of labor: agri-food complexes and export agriculture', in Friedland, W., Busch, L., Buttel, F. and Rudy, A. (eds), *Towards a New Political Economy of Agriculture Westview,* Boulder, Colorado, pp. 65-93.

Friedmann, H. (1993), 'The regulation of international markets: the unresolved tensions between nation states and transnational accumulation', Institute of Development Studies Bulletin, vol. 24, no. 3, pp. 49-53.

Friedmann, H. and McMichael, P. (1989), 'Agriculture and the state system. The rise and decline of national agricultures, 1870 to the present', *Sociologia Ruralis*, vol. 29, pp. 97-113.

Gereffi, G. (1994), 'The organization of buyer-driven global commodity chains: How U.S. retailers shape overseas production networks', in Gereffi, G. and Korzeniewicz, M. *(eds), Commodity Chains and Global Capitalism,* Greenwood Press, Westport, CT, pp. 95-122.

Gereffi, G., Korzeniewicz M. and Korzeniewicz R. (1994), 'Introduction: global commodity systems', in Gereffi, G. and Korzeniewicz, M. *(eds) Commodity Chains and Global Capitalism,* Greenwood Press, Westport, CT, pp. 1-14.

Goodman, D., Sorj B. and Wilkinson, J. (1987), *From Farming to Biotechnology,* Blackwell, Oxford.

Hill, S. (1991), 'How do you manage the flexible firm? The total quality model', *Work, Employment and Society,* vol. 5, no.3, pp. 397-415.

Jessop, B. (1990), 'Regulation theories in retrospect and prospect', *Economy and Society,* vol. 19, pp. 153-216.

Le Heron, R. (1988), 'State, economy and crisis in New Zealand in the 1980's: implications for land-based production of a new mode of regulation', *Applied Geography*, vol. 8, pp. 273-290.

Le Heron, R. (1993), *Globalised Agriculture. Political Choice,* Pergamon, Oxford.

Le Heron, R. and Park, S. (eds) (1995), *The Asian Pacific Rim and Globalization,* Avebury, Aldershot.

Le Heron, R. and Pawson E. (eds) (1996), *Changing Places, New Zealand in the Nineties,* Longman Paul, Auckland.

Le Heron, R. and Roche, M. (1995), 'A fresh place in food's space', *Area,* vol. 27, no. 1, pp. 22-32.

Le Heron, R. and Roche, M. (1996), 'Globalisation, sustainability and apple orcharding, Hawkes Bay, New Zealand', *Economic Geography,* forthcoming.

Leyshon, A. (1992), 'The transformation of regulatory order: regulating the global economy and environment', *Geoforum, vol.* 23, no. 3, pp. 249-267.

Lipetz, A. (1986) 'New tendencies in the international division of labour: regimes of accumulation and modes of regulation', in Scott, A. and Storper, M. (eds) *Production, Work, Territory: The Geographical Anatomy of Industrial Capitalism,* Allen, and Unwin, Boston, 16-40.

Long, N. (1995), 'Commoditization, livelihoods and contests of value. An actor perspective on the valorization and globalization of food', paper presented to the Political Economy of Agro-food Systems Workshop, University of California Berkeley, Berkeley, September.

Lynch, B. (1996) 'The meat industry', in Le Heron R., and Pawson E. (eds), *Changing Places. New Zealand in the Nineties,* Longman Paul, Auckland.

McMichael, P. (ed.) (1994), *The Global Restructuring of Agro-Food Systems,* Cornell University Press, Ithaca, New York.

Parsons, H. (1995), The Role of the Melbourne Wholesale and Vegetable Market in Fresh Produce Supply Chains, Working Paper No. Environmental Science, MonashUniversity, Melbourne.

Perry, M. (1995), 'Industry structures, networks and joint action groups', *Regional Studies*, vol. 29, no. 2, pp. 181-217.

Perry. M. and Goldfinch, S, (1996), 'Small business networking outside the industrial district', *Tidjschrift voor Economische en Sociale,* in press.

Perry, M., Davidson C. and Hill R. (1995), *Reform at Work*, Longman Paul, Auckland.

Porter, M. (1990), *The Competitive Advantage of Nations,* The Free Press, New York.

Sayer, A. (1989), 'Postfordism in question', *International Journal of Urban and Regional Research,* vol. 13, no. 4, pp. 666-696.

Storper, M. (1992), 'The limits to globalization: technology districts and international trade', *Economic Geography,* vol. 68, no. 1, pp. 60-93

Walker, R. (1988), 'The geographical organisation of production systems', *Environment and Planning D: Society and Space,* vol. 6, pp. 377-408.

Wallerstein, I. (1991), *Geopolitics and Geoculture: Essays on the Changing World-System,* Cambridge University Press, Cambridge.

5 Strategic alliances in global competition: securing or gaining the competitive edge

Wolf Gaebe

Introduction

In investment intensive and technology intensive sectors, not only has vertical cooperation by externalisation increased strongly since the 1980s, but so has cooperation between competitors in areas such as the development of new technologies and materials, the production of parts and distribution. While hierarchical relationships in and between business organisations are being reduced, new cooperative competition strategies are becoming increasingly significant, especially strategic alliances (Amin and Thrift, 1992, p. 575). It is not just industrial firms that enter into strategic alliances but increasingly service firms, such as banks and airlines. The term 'strategic alliance' has been in widespread use since the early 1980s and reports of new strategic alliances are common in the business literature. Specific information on these alliances is rather fragmentary and incomplete, and there is little detail on official and unofficial agreements and different forms of cooperation. Strategic alliances are a response to the Fordist structural crisis and intensifying global competition and they can be interpreted as being used by firms to stabilise, improve or regain their competitive edge.

The crisis in Fordism and the need for cooperation

It is important from the outset to recognise that the radical structural changes that have occurred in production and consumption in the past 25 years have brought great instability to the world economy. The crisis in Fordism that began to develop in the 1970s has been recognised as involving at least 10 distinct elements (Ekinsmyth et al, 1995; Peck and Tickell, 1994; Hirst and Zeitlin, 1992):

- the rise of flexible accumulation and flexible specialisation replacing the rigidities of Fordism and Taylorist organisation;
- the creation of new spatial divisions of labour;
- the emergence of new clusters of economic activities, both artisan and hi-tech based;
- a quickening of the pace of technological change in production, servicing, management and distribution;
- the growth of new key sectors of production and services;
- the energence of new methods of enterprise organisation;
- the emergence of new financial instruments and the commodification of money;
- the internationalisation and globalisation of production;
- the emergence of new inter-organisational relationships based loosely on cooperation; and
- the speeding up of the circuits of capital.

The crisis has generated incompatibilities between production and consumption structures that are only too evident in current high levels of unemployment and the falling competitiveness of firms. Markets are no longer stable and have become hard-to-control buyers' markets made up of discriminating, quality conscious consumers. Attempts by national governments to stabilise their economies using interventionist Keynesian policies have failed in the face of internationalisation. Old systems of regulation embracing existing systems of 'real regulation', can no longer stabilise the production-consumption exchange process. A return to and re-emphasising of core competencies is increasingly becoming a key criterion in the reorganisation of firms. The conglomerate corporations, such as Daimler-Benz, that were created in the 1970s and 1980s through diversification to spread risk have proved costly and difficult to run effectively.

However, concentration on core competencies and associated down-sizing are not efficient if they restrict a firm's growth. Under these circumstances, cooperation between business enterprises is an obvious coping strategy and strategic alliances are an important form of cooperation. Cooperation among suppliers in high technology sectors is becoming increasingly strong, and vertical disintegration to achieve flexibility is a clear spur to cooperation. Cooperation also allows firms to internationalise by overcoming barriers to entry, especially those barriers involving local content requirements. New communications and information technologies increasingly facilitate cooperation by allowing the central steering, co-ordination and monitoring of research, development, production, marketing and distribution, and the control of finance and investment. Economic instability can, therefore, be seen as the root cause of the search by business enterprises for cooperation and strategic alliances. Indeed, as Dertouzos et al (1989) have argued from studies of business enterprises in North America, Europe and Japan,

Undeveloped cooperative relationships between individuals and between organizations stand out ... as obstacles to technological innovation and the improvement of industrial performance. (p.7)

The following sections of this chapter will first define what is meant by a 'strategic alliance', then explore the advantages that firms can expect to achieve through their participation in such cooperative arrangements. Finally, the chapter will examine the impacts and significance of strategic alliances including their spatial and geographical impacts on patterns of production and investment.

Defining 'strategic alliances'

The definitions of strategic alliances in the literature of inter-firm cooperation are contradictory and mostly arbitrary as too are the evaluations of those alliances and the conclusions that have been drawn about them (see Rotering and Burger, 1994, p.109-111). In this chapter, strategic alliances are defined as:

> *problem-related cooperation between legally and economically (and functionally) independent firms.*

Through cooperation, competitive advantages, cost savings and strategic advantages are sought. Strategic alliances usually refer to cooperation between competitors in a particular sphere of business, for example in research and development, in production, or in distribution (see Table 5.1). They do not encompass the entire value added chain and all the functions of the cooperating partners' businesses, only a limited number of value added elements and functions. In all other areas of business the cooperating firms may remain competitors. Thus, strategic alliances represent a combination of cooperation and competition between legally independent firms although they can, at the same time, severely limit the economic independence of the firms involved. Economic independence is retained because each partner can unilaterally terminate the cooperation, and strategic alliances are abandoned when the aims of cooperation have been achieved, when a partner is no longer satisfied with the arrangement, or when company strategies are revised. Technological change and market shifts, for example, may cause common interests to disappear and conflicts to develop.

Against this definition and background, the planned merger between Renault and Volvo which was used by Alvstam (1995, p.44) as an example of a strategic alliance, can not be considered to be such an alliance because Volvo would have lost its independence if the merger had been completed. Nevertheless, a strategic alliance does describe an agreement to produce products jointly.

Table 5.1
Characteristics of strategic alliances

- Cooperation of firms in a specific functional range, e.g. research and development, production or distribution, on the basis of agreements (mostly horizontal cooperation).

- Legally independent cooperation partners.

- Economic independence in the ranges included in the cooperation.

- Various formal organisation of cooperation (contracts of corporate law, cooperation contracts, production agreements, distribution arrangements).

- Cooperation without capital invested (often limited as to time) or with capital invested (participation, joint ventures).

- Open and hidden cooperation aims (strategic competitive advantages).

- Expedient forms of cooperation between selfish partners, delicate balance between cooperation and competition.

- Spatial integration (concentration) or disintegration (de-concentration).

The structures and legal forms of strategic alliances can, in fact, be many and varied. Normally they include agreements on aims, means and also possible formal and informal sanctions. However, cooperative agreements between partners may not been be formulated precisely or completely. Profit sharing is common when alliances involve the sharing of costs and risks, but when partners pursue different aims through an alliance they often agree to share returns equally. In some circumstances it is not possible to express the returns from cooperation in exact monetary terms and this is very much the case with the shared returns from joint research and development.

The problems of joint problem solving and of the legal and economic independence of firms involved in alliances apply not only to cooperation between competitors (horizontal cooperation) and to firms in different sectors (diagonal cooperation), but also to the development partnerships between suppliers and customers (vertical cooperation). In most studies of strategic alliances there is no agreement about whether vertical cooperation should be considered as a specific type of alliance. However, it would seem unnecessarily restrictive to maintain that alliances can only be established between partners in the same line of business. By combining strengths in different, but complementary, lines of business, it is possible for strategic alliances to achieve particularly strong synergies.

It would also appear that different forms of cooperation are established to achieve different goals. While horizontal cooperation may aim primarily at the reduction of risk, vertical cooperation between suppliers and customers may to be improve efficiency. Supply agreements, franchising, licensing and concession agreements are not strategic alliances, and neither are mergers and majority ownership participation. In the case of supply, franchise, license and concession agreements, it is not cooperation which is agreed on but the performance of a contract at an agreed price - in effect a market transaction. In the case of majority ownership participation and mergers, one partner loses legal and economic independence.

Typically, strategic alliances involve:

- cooperation between large firms;
- an international orientation; and
- cooperation limited to a specific period of time.

Strategic alliances exist between large and small firms as well as between small firms though, typically, small and medium sized businesses are afraid of losing their independence. Between small firms, network relationships are more common than strategic alliances. However, exclusively national alliances have been formed and also unlimited, long-term forms of cooperation. Table 5.2 lists examples of strategic alliances that have been developed in the electronics industry in both research and development and in production.

Table 5.2
Examples of strategic alliances in the electronics industry

R&D
- Joint development of the PowerPC microprocessor by Apple (California), IBM (New York) and Motorola (Illinois) in the US.
- Joint development of a new computer chip by IBM, Siemens and Toshiba in the US.

Production
- Joint manufacturing of household appliances by Bosch-Siemens (Munich) and Wuxi Little Swan (China) in a joint venture in China.
- Joint manufacturing of programmable controls by AEG and Schneider, the French electronics group, in a joint venture in Bavaria.

Strategic alliances and competitive advantage

Strategic alliances allow a number of competitive advantages to be created and realized by firms. Advantages that can be realized directly by firms include, synergy effects and cost savings, time saving advantages, learning effects, risk reduction and the maintenance and imposition of standards. In addition to maximising the direct advantages gained from a partner's knowledge or know how, less obvious indirect advantages can be gained from strategic learning at a partner's expense, insights into a partner's organisation and strategies, and the prevention of a partner cooperating with other companies.

Synergic and cost advantages can be realised in strategic alliances by linking partners' complementary strengths, and by creating a critical mass for procurement, production or sales, or through development partnerships or access to a partner's distribution network. Manufacturers of components, for example, who want to become system suppliers can use strategic alliances to achieve economies of scale and to increase value added. Competition amongst producers of such products is mainly on price. With the help of an alliance it is possible for a firm to achieve a critical mass faster than on its own so that, in the face of sharply increasing costs for research and development and ever shortening innovation and product life cycles (made worse by new technologies increasingly involving large jumps), unit costs can be reduced. Strategic alliances are also an important instrument for external growth, especially for small firms, making it possible for them to achieve critical production volumes faster, more cost effectively and with less risk than on their own. Firms from high wage countries can achieve substantial cost saving through development partnerships, for software for example with counterparts in India, Taiwan or Russia where there are well educated, creative and motivated development groups.

Strategic alliances can also protect and enhance market access. Cooperation can help firms to adapt their products to markets and this has been particularly important for entry into South East Asian markets. According to commercial surveys, entry into new markets is the main aim of strategic alliances, ahead of cost reduction and product range expansion. This functional aspect of strategic alliances is clearly evident, for example, in the marketing strategies of the European suppliers of high speed trains. Siemens, the leader in the ICE consortium and GEC Alsthom, the British-French manufacturer of the TGV, market together in Asia but separately in Europe and North America. In Asia they do not compete with one another because it could result in prices that would not cover costs. In Europe and North America, however, GEC Alsthom has greater market penetration and does not need to cooperate with Siemens.

Time savings advantages arise from strategic alliances because the shortening product development periods is increasingly necessary for commercial success (see Dicken et al, 1994, p. 31-32). Not only can transaction costs be reduced, but also time can be gained by exchanging knowledge and experience in strategic alliances. Often it is only the initial suppliers to a market, the pioneers, who are able to amortise their large investments. Cooperation in research and development has

106

become a fixed element of firm strategies. Instead of striving for technological self-sufficiency, companies now seek strategically to use technological expertise. Cooperation has become a means of sharing the high costs and risks of innovation and of satisfying the need for knowledge and quality. In these circumstances, firms use strategic alliances whenever the exercise of influence or takeover are either impossible or not expedient. It would appear that R&D joint ventures, joint R&D projects and direct investment are for the most part strategically motivated, whereas technology exchange agreements and customer-supplier alliances tend to be orientated more towards cutting costs (Hagedoorn, 1993, p. 128). Firms from less developed countries, especially the newly industrialising countries, see quick access to new technology as the main advantage of strategic alliances and this is why South Korean firms like Samsung have entered into strategic alliances with Japanese firms.

Cooperation and marketing, mostly between foreign and domestic firms, has existed for a long time, but cooperation to gather or to exchange knowledge is relatively new. Many firms feel that patents and registered designs are insufficient to protect their know-how or technological advantages. Vertical and horizontal alliances, cooperation with customers and suppliers together with less frequent cooperation with competitors, are important for the innovation process and as sources of innovation in their own right. The gaining of knowledge from less accessible sources is a strong motivation to form strategic alliances. However, information, knowledge and know-how are often difficult to put monetary values on and the results of technical studies can be uncertain and difficult to foresee. It is, therefore, difficult to formulate in a legally precise and realisable manner the performance that can be expected from any strategic alliance. Nethertheless, because complex knowledge can be turned into traded commodities, alliances are becoming increasingly technology orientated, with cooperating firms looking more and more for access to new processes and research methods (Storper, 1992, p. 82).

The partners in strategic alliances will make efforts to learn as much as they can and to internalise background knowledge and know-how, though R&D cooperation requires knowledge inputs from all partners. Through strategic alliances, knowledge can be kept out of the public domain and especially away from competitors. Cooperation means, therefore, that partners can hang on longer to a technological lead they may have. Importantly, cooperation also means that duplicated and parallel research can be avoided leading to the better allocation of scarce resources in the creation of new knowledge.

The cost of R&D for new products can also be reduced through the cooperation of strategic alliances. R&D costs, especially in tele-communications and pharmaceuticals, have increased to such an extent that even very large firms can not bear all the costs and risks alone. Many firms cooperate with competitors because they are not certain that an investment will be successful. Examples of strategic alliances to reduce costs and risks and gain time are the cooperation between the Japanese NEC corporation and the South Korean Samsung Group to develop the 256-megabit chip, and the cooperation between IBM, Motorola, Siemens and Toshiba to develop at a cost of $US 1.3 billion gigachips with four times the memory of a 256-megabit chip. Another example is the joint development of turbines for the planned Boeing Super Jumbo by General Electric and Pratt & Whitney. Here strategic alliance is necessary because of the estimated $US 1.5 billion development costs and the risks associated with developing an aircraft that is still only in the planning stages.

In addition to legally formulated standards, strategic alliances play an import role in determining de facto, unofficial standards for products - generally accepted products characteristics. '[S]tandards agreements form a very small proportion of the whole population of alliances .. but they are probably the most important type in terms of global impact' (Cooke and Wells, 1991, p.351). An example of an alliance's product solution that became a de facto industry standard is the VHS video system of the Japanese Matsushita Group (Panasonic, Technics and JVC) which gained market acceptance over the competing Philips and Grundig system. In contrast to its European competitors, the Japanese alliance had secured the early expansion of its system by licensing. As a result, Philips and Grundig abandoned their system, though it was considered superior in quality, and began to produce VHS recorders under license. Other examples of this creation of unofficial industry standards can be seen in the alliance of Sony, Matsushita Electric, Toshiba and Philips to produce digital video compact discs, and the alliance of Kodak, Fuji, Canon, Minolta and Nikon to develop the new Advantix camera system. Today, de facto, unofficial standards play a decisive role in determining success in the market: those who create the industry standard have the market. High levels of market penetration help to turn a product into an industry standard. These critial levels of penetration can be achieved much more quickly when firms cooperate rather than act alone.

In general, however, strategic alliances are used to pursue several objectives at the same time: to gain time and learning advantages and to reduce and spread risk, for example (Servatius, 1990, p. 58). Strategic alliances in research and development can represent a preliminary stage in the formulation of production and distribution agreements, or they can be part of such agreements. Equally, they may constitute the initial steps towards a takeover or merger. The number of strategic alliances in biotechnology, for example, between large multinational companies and relatively small, specialised, R&D-intensive companies has decreased substantially in recent years because, besides economic problems and bankruptcies, many small firms have been taken over by their former partners (Hagedoorn, 1993, p. 122).

Strategic alliances are expedient forms of cooperation between selfish partners. Because they involve obvious disadvantages, such as the early disclosure of information on know-how and the internal organisation of partner firms, they are often only entered into when an internationalising company needs a domestic partner in an overseas venture (see Ohmae, 1989, p. 143; Lundvall, 1993, p. 57). Those alliances last only as long as the advantages of cooperation outweigh the disadvantages:

> ... partnerships remain unlikely as the basis for stable long-term success in the face of strong competition ...Attractive as they are, and while these

108

alliances indeed provide an opportunity for fast global market access, they may not allow the degree of strategic control, change and flexibility which would be needed to secure long-term competitiveness. (Doz et al, 1990, p. 117, 122)

Medium-sized firms are often not prepared to cooperate with competitors, even when they are under considerable political pressure to do so. The establishment of long-term relationships between firms, though commonly cited in the literature, may often be quite unintended. If there were no political or economic impediments or risks, owning a firm overseas would be preferable to most business enterprises that alliances in the form of joint ventures. As Barnevik, CEO of ABB (Asea Brown Boveri) has remarked, 'The strategic alliance with Daimler-Benz was advantageous in railroad technology, in other areas ABB has to live with competition, new global strategic alliances are unnecessary.' (*Financial Times,* April 1996)

Many firms are in transition from being exporters to being global competitors, from being technological followers to being technological pioneers, and from conducting research in their home countries to researching in international networks. They expect that strategic alliances at this stage in their development will enable them to overcome their current problems. Strategic alliances are thought to make access to markets, innovations and international research easier.

Cooperation between firms has become more intensive as international competition has become more fierce, and technology has become more important as the basis of competitive strength in sectors requiring a high level of research and development. This apparently contradictory statement can be explained by firms choosing strategies that combine both cooperation and competition. For example, they cooperate in the research on and development of new products, but they compete in the distribution of those products. Firms can cooperate in a national or international alliance, but at the same time compete with other strategic alliances. The complexity of possible arrangements is demonstrated by strategic alliances in the aero engine construction industry. In April 1990, General Electric sued Daimler-Benz for more than US$ 1 billion in damages because, MTU, a Daimler-Benz subsidiary, had agreed to cooperate closely with the rival Pratt & Whitney, although an alliance already existed with General Electric. In May 1990, Daimler-Benz and General Electric, settled the suit out of court.

The impact and significance of strategic alliances

The formation of strategic alliances between firms has had a major impact on inter-firm cooperation and the operations and functioning of large firms as they struggle with processes of internationalisation and the globalisation of competition. Four main impacts of these strategic alliances can be identified:

- growth in the number of alliances leading to the establishment of global networks of commercial alliance;
- the compounding of impacts on competition, with alliances created in reaction to competitive pressures then impacting directly and indirectly on competition;
- the fine line that exists between strategic alliances and restrictive cartels;
- the differential benefits alliances bestow, often serving partners well but not necessarily their customers.

The number of strategic alliances has increased very substantially in recent years, and it is primarily firms from the emerging triad of trading blocs (Europe, North America and Asia) that participate in these cooperative arrangements (Dicken, 1992). Complex networks of alliances between competitors now span whole branches of industry, for example the automobile industry, consumer electronics and telecommunications. US, European and Japanese firms have concluded more cooperative agreements with foreign rather than domestic firms, and strategic alliances are most commonly established for the development of new technologies and new materials.

The OECD (1992) has outlined the forms of cooperation between firms in relation to science and technology, and identified alliances to expand technological bases, improve competitive strengths, enhance revenues and to expand sales. The Centre for European Economic Research (ZEW) in Mannheim estimated from panel research that about half of all firms who carry out R&D undertake joint R&D projects through alliances with other firms or with scientific establishments (Licht, 1994, p. 371).

Strategic alliances established to react to competition in turn tend to go on to further influence competition. Alliances between large firms to create informal, de facto standards have the potential fundamentally to change competition. However, competition can also be intensified when such alliances are formed because partners may be introduced to new markets for the first time.

However, in terms of competition, a distinction must be made between, on the one hand, economically rational cooperation through the creation of strategic alliances and, on the other hand, the creation of restrictive cartels. Global industrial cartels can evolve as a result of cooperation and the harmonisation of interest between firms, and they have the potential to restrict trade owing to the power the alliance can wield. However, strategic alliances also carry considerable risk for the partners themselves since the partner who is also a rival can again access to new know-how within the alliance and can thus establish a low-cost bridgehead into new markets. Indeed, Porter (1990) has gone so far as to conclude that '... alliances as a broad-based strategy will only ensure a company's mediocrity, not its international leadership. No company can rely on another outside, independent company for skills and assets that are central to its competitive advantage.' (p. 93)

The evidence of strategic alliances in civil aviation shows that inter-firm cooperation can bring advantages to partners but disadvantages to customers. A world-wide network of airline alliances is developing rapidly, partly in response to markets distorted by regulation and subsidy. In 1996, there were approximately 280 strategic alliances among airlines. These airline networks all have much the same pattern - each includes one North American, one South American, one European and one Asian airline. Lufthansa, for example cooperates with Air Canada (since 1996), United Airlines (since 1995), Varig (since 1993), SAS (since 1995), Adria Airways (since 1995), South African Airways (since 1995) and Thai International Airways (since 1994). The advantages for the airlines are obvious: reduced overcapacity and better capacity utilisation, lower advertising and marketing costs, and access to expanded route networks. The disadvantages for passengers are lack of knowledge about who ultimately will be their carrier, higher prices, less competition, and greater difficulty in transferring between airline networks.

In micro-electronics, strategic alliances are one of a range of strategies being used by firms to pursue goals of internationalisation. In this industry, internationalisation is achieved through the establishment of new subsidiaries and participation in joint ventures as well as through strategic alliances. Siemens, for example, which is the largest European electrical engineering group, operates in a great many locations, has numerous subsidiaries and is engaged in alliances with other firms in virtually every continent. Amongst these alliances, Siemens cooperates, for example with Sony for the manufacture of mobile telephones and Motorola and Toshiba for the production of semi-conductors.

The spatial significance of strategic alliances

Studies of strategic alliances mainly concentrate on the characteristics, goals, and significance of this form of cooperation and on the problems of the choice of partner. Even in empirical studies very little has been written about the spatial effects of strategic alliances, for example about the effects on international integration, direct investment, jobs, research, and the effectiveness of government regulatory systems (see Gugler, 1991). It is difficult quantitatively to assess the significance of strategic alliances because few details are available on numbers of jobs, value added, market share, and the turnover of these joint projects. Qualitative statements about spatial effects, for example about the growth of knowledge and competence, are even more difficult. Indeed, '[u]nlike acquisitions, there are no comprehensive statistics on the incidence of strategic alliances, and estimates ...vary considerably' (Hamill, 1993, p. 104). Another reason why questions about the spatial effects of strategic alliances are difficult to answer is that '... in most of the case study firms ... the rise in alliance activity ... was only part of a wider corporate restructuring process which involved ... the whole range of corporate activity both in terms of internal and external activities.' (Cooke and Wells, 1991, p. 350).

The effects of strategic alliances on jobs depend on the aims of the cooperative venture. If the aims are to open up markets and to protect them, employment is likely to be increased. Job losses resulting from rationalisation of the value added chain can

111

be offset by new jobs resulting from increased competitiveness and new investment. New investment can also create new jobs which offset losses resulting from relocation and integration in the value added chain ('flexible integration', Amin, 1993, p. 290) when the aims of strategic alliances are focused on time savings, learning effects, and the creation of de facto industry standards. Such investment mainly takes place in existing locations and within the partners' networks. If, however, the goals are cost reduction, greater efficiency, and synergy effects, there is more likely to be a reduction, and not an increase, in employment and job security as a result of rationalisation and the integration of the alliance into the value added chains of one or both partners. Many alliances occur in fields of business in which a large part of the synergy potential is in markets which are very limited in spatial terms (Abravanel and Ernst, 1994, p. 269). An empirical investigation of strategic alliances has confirmed the hypothesis that strategic alliances, whose primary aim is to achieve production cost advantages, tend to occur within Taylor-type forms of organisation and thus help bring about a high degree of polarisation of the labour market or increase an already existing polarisation.

Strategic alliances bring about increases in international relationships, direct investment, and research in technologically advanced countries. The effects on competition are less clear. Competition can become more intense as a result of strategic alliances but it can also become more limited. In many branches, the general setting has been changed by strategic alliances, for example, for suppliers of transport, telecommunication and personal services and software. Global alliances compete with one another here. Thus, international air traffic is increasingly dominated by a few large alliances. For example, a quarter of all flights between Europe and the USA, and about 60 per cent of those between Great Britain and the USA, are made by the partner airlines British Airways and American Airlines. Private market arrangements of this type can reduce competition, for example through regional monopolies. Any competition among airlines mostly occurs outside the protected domestic market. Long distance flights, for example from Germany by way of London, are considerably cheaper than flying directly from London. It seems likely that feeder services from the continent to London will increase as a result of the strategic alliance between British Airways and American Airlines.

In telecommunications, too, alliances are increasingly being used to provide services. For example, British Telecom and MIC's joint venture - the speech and data 'concert' - is providing services in 60 countries.

Strategic alliances between leading large firms in oligopolistic mass markets increase competition in the short run. In the longer run, however, they restrict competition because potential competitors, particularly medium sized or small firms, are frightened off.

Transnational networks involving a comprehensive exchange of information and services have also been established in R&D. International research networks and research in the technologically advanced countries are increasing. Research concentrated in a single location is falling in importance. Japanese firms, for example, prefer to set up research laboratories in regions with leading international research establishments because of the expected learning effects. They have set up laboratories in Princeton, Boston, and the New York area in the USA and in selected places in Europe.

Many markets, for example in China, can only be entered by means of strategic alliances between host country ventures and foreign partners. Others are difficult to enter without such alliances because of local content requirements. In most cases only a few strategic alliances are allowed so that competition is extremely limited. However, competition can become stronger if the market power of the partner is strengthened, or if the strategic alliance allows a firm to enter a particular market for the first time. Competition can also increase if factors of production which were formerly tied to a particular location become widespread - for example, complex knowledge about new products and processes that is integrated into regional relationships and into a regional milieu. A competitive situation can often develop with a strategic alliance between strong partners from the same triad. This competitive aspect prevents firms gaining that access to markets and competences that they could otherwise ensure by setting up an alliance with strong partners (Abravanel and Ernst, 1994, p. 280).

Strategic alliances make it possible for firms to circumvent government controls and systems of regulation. But, much depends on whether a TNC is dealing with its country of origin as a home-base or whether it is investing in a host to which it owes no allegiance. Strategic alliances can help to loosen these ties (Harrison, 1994, p. 223; Dicken et al, 1994). Strategic alliances with an advanced country only give an impetus to development, through for example the transfer of technology when a partner is interested in a long term commitment (Bleeke and Ernst, 1994, p. 39).

Conclusion

It is not only vertical cooperation through externalisation that has increased greatly in investment and technology-intensive branches of economic activity that has increased since the 1980s but also cooperation between competitors in the form of strategic alliances. Strategic alliances are a reaction by firms to the shift away from Fordist structures of production and consumption, the intensification of in global competition together with the internationalisation and globalisation of economic activity. They provide firms and business organisations with an alternative route by which to achieve growth. They have come to prominence in situations where going it alone does not seem sensible, where taking share holdings in other businesses or merging with them are not possible, and when knowledge is not tradeable and can only be acquired through cooperation with a partner. Many overt and covert aims are pursued by means of strategic alliances - aims which have become very important for achieving increased competitiveness. These aims include ensuring and opening up markets, the achievement of synergy effects, time savings, learning effects, risk reduction, and the establishment of de facto industry standards. Normally strategic alliances involve only one field of business or one function, for example R&D.

Strategic alliances are, in fact, goal directed forms of cooperation between self-interested actors and, because of their obvious disadvantages, are only entered into because a partner is necessary. They last only as long, as the advantages persist. A strategic alliance can turn out to be a Trojan horse, particularly when the strengths of the partners are very different. Most alliances end with one partner buying the other partner's share. Other alliances end in separation instigated by the stronger, and better placed partner. The empirical evidence shows that strategic alliances, especially alliances with local partners in global business, appear primarily to be advantageous when new fields of business or new markets are opened up. They are less advantageous when the aim is to strengthen core business activities. In the latter case, mergers and acquisitions are more effective than strategic alliances (Bleeke and Ernst, 1994, p. 35).

References

Abravanel, R. and Ernst, D. (1994), 'Allianz versus akquisition: strategische optionen für europäische "Landesmeister"', in Bleeke, J. and Ernst, D. (eds), *Rivalenals Partner. Strategische Allianzen und Akquisitionen im globalen Markt*, Frankfurt and New York, pp. 269-291.

Alvstam, C.G. (1995), 'Spatial dimensions of alliances and other strategic manoeuvres', in S. Conti, S., Malecki, E. and Oinas, P. (eds), *The Industrial Enterprise and its Environment: Spatial Perspectives*, Avebury, Aldershot, pp. 43-55.

Amin, A. (1993), 'The globalization of the economy. an erosion of regional networks?', in Grabher, G. (ed.), *The Embedded Firm. On the Socioeconomics of Industrial Networks*, Routledge, London, pp. 278-295.

Amin, A. and Thrift, N. (1992), 'Neo-marshallian nodes in global networks?', *International Journal of Urban and Regional Research*, vol. 16, pp. 571-587.

Bleeke, J. and Ernst, D. (1994), 'Mit internationalen allianzen auf die siegerstraße', in Bleeke, J. and Ernst, D. (eds), *Rivalen als Partner. Strategische Allianzen und Akquisitionen im globalen Markt,* Frankfurt and New York, pp. 34-53.

Cooke, P. and Wells, P. (1991), 'Uneasy alliances: the spatial development of computing and communication markets', *Regional Studies*, vol. 25, no. 4, pp. 345-354.

Dertouzos, M.L., Lester, R.L. and Solow, R.M. (1989), *Made in America. Regaining the Productive Edge*, Cambridge, MA.

Dicken, P. (1992), *Global Shift: The Internationalisation of Economic Activity*, 2nd edition, Paul Chapman Publishing, London.

Dicken, P., Forsgren, M. and Malmberg, A. (1994), 'The local embeddedness of transnational corporations', in Amin, A. and Thrift, N. (eds), *Globalization, Institutions, and Regional Development in Europe*, Oxford University Press, Oxford, pp. 23-45.

Doz, Y., Prahalad, C.K. and Hamel,G. (1990), 'Control, change and flexibility: the dilemma of transnational colloboration', in Barlett, C., Doz, Y. and Hedlund, G. (eds), *Managing the Global Firm*, London, pp. 117-143.

Ekinsmyth, C., Hallsworth, A.G., Leonard, S. and Taylor, M. (1995), 'Stability and instability: the uncertainty of economic geography', *Area*, vol. 27, pp.289-299.

Gugler, P. (1991), *Les Alliances Stratégiques Transnationales*, Fribourg, Documents Economiques 58.

Hagedoorn, J. (1993), 'Strategic technology alliances and modes of cooperation in high-technology industries', in Grabher, G. (ed.), *The Embedded Firm. On the Socioeconomics of Industrial Networks*, Routledge, London, pp. 116-137.

115

Hamill, J. (1993), 'Cross-border mergers, acquisitions and strategic alliances', in Bailey, P., Parisotto, A. and Renshaw G. (eds), *Multinationals and Employment. The Global Economy of the 1990s.* Genf, pp. 95-123.

Harrison, B. (1994), *Lean and Mean. The Changing Landscape of Corporate Power in the Age of Flexibility*, New York.

Hirst, P. and Zeitlin, J. (1992), 'Flexible specialization versus post-fordism: theory, evidence, and policy implications', in Storper, M. and Scott, A.J. (eds), *Pathways to Industrialization and Regional Development.* Routledge, London, pp. 70-115.

Licht, G. (1994), 'Gemeinsam forschen. Motive und verbreitung strategischer allianzen in Europa', *ZEW Wirtschaftsanalysen*, vol. 2, pp. 371-399.

Lipietz, A. (1986), 'New tendencies in the international division of labor: regimes of accumulation and modes of regulation', in Scott, A.J. and Storper, M. (eds), *Production, Work and Territory. The Geographical Anatomy of Industrial Capitalism*, Allen and Unwin, London, pp. 1-40.

Lundvall, B. (1993), 'Explaining interfirm cooperation and innovation: limits of the transaction-cost approach', in Grabher, G. (ed), *The Embedded Firm. On the Socioeconomics of Industrial Networks*, Routledge, London, pp. 52-64.

OECD (1992), *Wissenschafts- und Technologiepolitik. Bilanz und Ausblick 1991*, Paris.

Ohmae, K. (1989), 'The global logic of strategic alliances', *Harvard Business Review*, vol. 67, no. 2, pp. 143-154.

Peck, J. and Tickell, A. (1994), 'Searching for a new institutional fix: the after-fordist crisis and the global-local disorder', in Amin, A. (ed), *Post-Fordism: A Reader*, Oxford University Press, Oxford, pp. 280-315.

Pfützer, S. (1995), *Strategische Allianzen in der Elektronikindustrie. Organisation und Standortstruktur*, Wirtschaftsgeographie Bd. 9, Münster, Hamburg.

Porter, M. E. (1990), 'The competitive advantage of nations', *Harvard Business Review*, vol. 68, pp. 73-93.

Rotering J. and Burger, C. (1994), 'Strategische allianzen zurs stärkung der wettbe werbsposition in der liberalisierung', in Booz, Allen & Hamilton (eds), *Gewinnen im Wettbewerb. Erfolgreiche Unternehmensführung in Zeiten der Liberalisierung*, Stuttgart, pp. 103-124.

Servatius, H.-G. (1990), 'Koordination internationaler strategischer allianzen', in Backhaus, K. and Piltz, K. (eds), *Strategische Allianzen. Eine neue Form kooperativen Wettbewerbs?*, Zfbf, Sonderh, pp. 49-66.

Storper, M. (1992), 'The limits of globalization: technology districts and international trade', *Economic Geography*, vol. 68, pp. 60-93.

The Economist Intelligence Unit (EIU) (ed) (1994), *Best Practices: Strategic Alliances*, EIU, New York.

116

Part II: Business enterprise, local networks and 'spatial learning'

Of primary concern in relation to increasing economic globalisation are the mechanisms available to maintain and retain economic autonomy and growth at the scale of the local community and the industrial district. The four chapters that make up this part of the volume address this aspect of the global-local dialectic from the perspective of locally embedded network systems. They provide four very different glimpses of growth, learning, coping and planning strategies to achieve and retain local economic autonomy. Malmberg and Sölvell examine the processes of spatial clustering and the local accumulation of knowledge as mechanisms creating local competitive advantage in industrial districts confronted by global competition. Asheim further refines this perspective through a very detailed appraisal of knowledge flows, learning, technology and innovation in industrial districts as 'learning regions'. Cuñat and Thomas review the socialisation of local network relationships within the old textile districts of northern France to elaborate a perspective on coping strategies through the creation of charters linking the powerful and the disempowered in an area severely disrupted by global restructuring pressures. Finally, Malecki and Tootle explore the for policy to generate local growth that is globally competitive through the creation of locally embedded business networks of SMEs in the USA.

6 Localised innovation processes and the sustainable competitive advantage of firms: a conceptual model

Anders Malmberg and Örjan Sölvell

Introduction

The influence of geographical space on economic performance has attracted interest in increasingly wide circles over the last decade such that not only geographers but also scholars in economics, technology and innovation and international business studies now address issues of space and place when studying the dynamics of industrial change (Storper, 1995). With the aim to improve our understanding of how firms sustain competitive advantage through continuous innovation over extended time periods, this chapter suggests a conceptual model built on knowledge emerging from three strands of research focusing on economic geography, technology and innovation, and international business, respectively.[1]

In economic geography there are two important concerns about location and firm performance. The first is that economic, entrepreneurial and technological activities tend to agglomerate at certain places, leading to patterns of national and regional specialisation (Malmberg and Maskell, 1996). The second is that the performance and development of a firm to a considerable extent seems to be determined by the conditions that prevail in its environment, and that the conditions in the immediate locality - in the local cluster or local milieu - seem to be particularly important (Saxenian, 1994; Scott, 1995).

An important part of the literature on technology deals with the innovation process, defined broadly to include anything from incremental improvements in product or production processes to major technological break-throughs (see Chapter 7 by Asheim in this volume). In current research on innovation and technological

119

development, there is a strong emphasis on the relation between the innovating firm and its external environment, including related firms, research bodies and other actors. An explicit interest in the role of the local milieu can often be seen in these approaches, even though empirical evidence predominantly equates this local arena with the nation when studying localised innovative interaction (Lundvall, 1992; Patel, 1995). Other research in the technology field has, through a long empirical tradition, now built a strong case for historically bounded patterns or trajectories locking in firms and nations over substantial periods of time (Dosi, 1982), again pointing towards location as a critical variable.

From one point of view, it might seem odd to bring in geographical concepts such as proximity and localisation as we build a model of the innovation process in an era of rapid globalisation. Indeed, as flows of people, products, information and capital are becoming increasingly extended in space (Dicken, 1992; Amin and Thrift, 1995), the question is if there remains a role to be played by the local environment. Some come to the conclusion that the role of the nation or region is rapidly diminishing with the emergence of global markets and firms, whereas others emphasise the new methods of production in the 'post-Fordist' era being a rejuvenating force for localisation (Piore and Sabel, 1984; Amin and Malmberg, 1992; Storper, 1995; Beccattini and Rullani, 1996). Business scholars focusing on the firm and the transnational corporation (TNC) in particular, have in a similar fashion conveyed doubts about the local environment as having any important role to play today in how global firms formulate their strategies and build competitive advantage (Prahalad and Doz, 1987; Bartlett and Ghoshal, 1990; Hedlund and Rolander, 1990; Ridderstråle, 1996). This has sparked a 'counter-reaction' from a group of scholars who argue instead that the local environment - labelled as the firm's home base - plays a continued or possibly increasingly important role for the global firm (Porter, 1990; Sölvell and Zander, 1995).

In this chapter, we will focus on a set of questions related to the impact of geographical location on a firm's ability to create and sustain competitiveness in an era of increased economic globalisation and, thus, increased exposure to international competition. What is the role played by agglomeration in general, and spatial clustering of related firms and industries in particular? The chapter brings together theory from economic geography, international business studies and innovation studies to address the phenomena of spatial clustering, accumulation of knowledge in local milieux and firm competitiveness. Here, we identify a coincidence of research interests, even though geographers typically have focused on the characteristics that determine a region's economic structure and performance (Storper and Walker, 1989; Storper, 1995), whereas international business strategists have focused on the role of regions and nations in shaping the competitive advantage of firms in general (Porter, 1990; Kogut, 1993) and transnational corporations in particular (Sölvell, Zander and Porter, 1991).

The aim of the chapter is to build an argument about why the accumulation of knowledge - essential as it is to firm competitiveness - involves important local elements tied to a region or nation, in spite of recent trends of international economic

120

integration and the enhanced international mobility of resources. The role of transnational corporations in linking together localised processes of innovation is discussed in this context.

The argument of the chapter is structured into three sections. The next section contains a discussion of theoretical antecedents, notably in relation to the agglomeration phenomenon as it has been dealt with in the economic and geographical literature. In the subsequent section, we turn to the process of local knowledge accumulation, emphasising the localised nature of the innovation process, the role played by barriers to knowledge diffusion, and the process whereby resources are attracted to a spatial cluster from the outside. Finally, we discuss how the innovation activities of transnational corporations link in with localised processes of knowledge creation.

Theoretical antecedents

A natural starting point for theoretical considerations regarding the existence of spatial clusters of related industries is the diverse body of academic work that can be assembled under the label 'agglomeration theory'.[2] It has evolved in response to three sets of empirical observations. The first is that a large proportion of total world output of manufactured goods is being produced in a limited number of highly concentrated industrial core regions. The second observation is that firms in particular industries, or firms which are related in other ways, tend to co-locate and thus form spatial clusters. A third observation is that both these phenomena tend to be persistent over time. Once in place, the agglomerative process tends to be cumulative, as Myrdal (1957), Hirschman (1958), Ullman (1958) and Pred (1977) had already noted several decades ago.

Agglomeration theory: static accounts

Traditionally, the long term survival of an established industrial region has been seen as resting on geographical inertia: a built-in resistance to decline that is sometimes extremely strong in preserving the location of an industry. This tendency of regions to sustain a given industrial specialisation, even if their original locational advantages have disappeared, has been attributed to the fact that capital equipment is more or less immobile, and that it is therefore often cheaper to expand industrial capacity at an existing site than to construct a new plant on a new site (Estall and Buchanan, 1961). Modern accounts of the durability of spatial clusters focus, as will be further discussed below, more on the learning abilities of local milieux.

The three sets of observations - we can call them regional concentration, spatial clustering and path dependence - have been described and analysed in some detail by numerous writers, from Marshall (1890/1916) and Weber (1909/1929) through Hoover (1948), Estall and Buchanan (1961) and Lloyd and Dicken (1977), to Krugman (1991a) and Enright (1994), to mention but a few. Often, a distinction is made between two types of agglomeration economies. One type relates to general economies of regional and urban concentration that apply to all firms and industries in a single location and represent those external economies passed on to firms as a result of savings from the large-scale operations of the agglomeration as a whole. These are the forces leading to the emergence of industrial core regions ('manufacturing belts'), large conurbations and metropolitan regions. A second type of agglomeration economies are the specific economies that relate to firms engaged in similar or inter-linked activities, leading to the emergence of spatial clusters of related firms ('industrial Hollywoods', 'new industrial districts', 'innovative milieux', etc.). Often, these two sets of forces are referred to as urbanisation economies and localisation economies, respectively (Lloyd and Dicken, 1977).

When it comes to the actual mechanisms that make up the agglomerative force, the distinction between the two is, however, less clear. In both cases, agglomeration economies have their roots in processes whereby links between firms, institutions and infrastructures within a geographic area give rise to economies of scale and scope: the development of general labour markets and pools of specialised skills; enhanced interaction between local suppliers and customers; shared infrastructure; and other localised externalities (Hoover, 1948; Lloyd and Dicken, 1977). Agglomeration economies are believed to arise when such links either lower the costs or increase the revenues (or both) of the firms taking part in the local exchange. Presence in an agglomeration is, in other words, held to improve performance by reducing the costs of transactions for both tangibles and intangibles (Appold, 1995). In this way, the key to agglomeration has been attributed to the minimisation of the distance between a firm and its trading partners, as well as to the speed with which communication can take place between customers and suppliers. In Scott's (1983; 1988) view, the formation of regionalised industrial systems will be particularly intense where linkages tend to be small-scale, unstable and unpredictable, and hence subject to high unit transaction costs.

The traditional accounts of the agglomeration phenomenon are predominantly static. It is first and foremost increased efficiency in transactions of goods and services that is believed to give rise to benefits for firms located in agglomerations. This strong focus on the efficiency and intensity of local transactions is somewhat paradoxical, since the much-theorised linkages between agglomerated firms has proven to be weak and indeed thin, in empirical studies (McCann, 1995). Thus, a large proportion of firms have few or no trading links with other local firms within the same industry, even when there is a strong spatial clustering of a particular industrial sector. Furthermore, a large proportion of firms have no or few trading links with other firms or households in the urban area or the geographical region where they are located, even though the area comprises a clustering of various economic activities. Such evidence would seem to make 'linkage-based' claims about the importance of agglomeration economies somewhat questionable.

Still, spatial clustering may well play an important role without any significant local input-output relationships. The rapid globalisation of markets and firms and increased competitive pressures in many industries has recently triggered a debate about competitiveness, both at the level of regions and nations and at the level of the firm. There is now broad agreement that the main parameters of competition cannot be fully captured within a traditional framework based on relative cost advantages. Sustained competitiveness has more to do with capabilities leading to dynamic improvement than with achieving static efficiency (Porter, 1990; 1994). Industrial systems are not just fixed flows of goods and services, rather they are dynamic arrangements based on knowledge creation (Patchell, 1993; Beccattini and Rullani, 1996).

In line with this new view of competitiveness, recent research approaches have come to focus on the importance of localised information flows and technological spill-over when trying to explain the emergence and sustainability of spatial clusters of related firms.[3] In this view, industrial systems are made up not only of physical flows of inputs and outputs, but also by intense exchange of business information, know-how, and technological expertise, both in traded and untraded form (Scott, 1995). In addition to reductions in the costs of inter-industry exchange and improved circulation of information and capital, increasing emphasis is now being placed on the claim that spatial clustering may lead to a reinforcement of transaction-based modes of social solidarity in a 'Marshallian atmosphere'. The latter point is of course far from new, since Marshall more than a century ago made a point of this 'social effect' of localisation when it comes to the promotion of upgrading:

> When an industry has thus chosen a locality for itself, it is likely to stay there long: so great are the advantages which people following the same skilled trade get from near neighbourhood to one another. The mysteries of the trade become no mysteries; but are as it were in the air, and children learn many of them unconsciously. Good work is rightly appreciated, inventions and improvements in machinery, in processes and the general organization of the business have their merits promptly discussed: if one man starts a new idea, it is taken up by others and combined with suggestions of their own; and thus it becomes the source of new ideas. And presently subsidiary trades grow up in the neighbourhood, supplying it with implements and materials. (Marshall 1890/1916, p. 271)

While Marshall's main concern was the existence and reproduction of spatial clusters of related firms,[4] there have been corresponding attempts to analyse the 'learning abilities' of regional and urban agglomerations more generally. In this context, Andersson (1985) has listed five preconditions that should be fulfilled if a region is to become and remain creative. It should have:

123

- high levels of competence;
- many fields of academic and cultural activity;
- good possibilities for internal and external communications;
- widely shared perceptions of unsatisfied needs; and
- a general situation of structural instability, promoting synergies.

Instead of specialisation and spatial clustering of related industries, emphasis is placed on the presence of a regional variety of skills and competencies, where the - often unplanned - interaction between different actors will lead to new - often unexpected - ideas (i.e. synergies). Malecki (1991) has reviewed several studies that, in much the same vein, define the properties of an entrepreneurial region (Johannisson, 1987).

Concepts of agglomeration forces

One way to order the various approaches to the agglomeration phenomenon is to make distinctions along two dimensions: between agglomeration forces operating at the general level or at the level of related firms and industries on the one hand, and forces increasing static efficiency or dynamic improvement on the other. In figure 1, we indicate how different notions of agglomeration could be fitted into such a matrix, and give examples of how various research approaches and scholars could be linked to these notions.

	Agglomeration of economic activity in general	**Spatial clustering of related firms and industries**
Transaction efficiency and flexibility	• **Manufacturing belts** (Ullman, Krugman) • **Metropolises** (Pred, Myrdal, Hirshman)	• **Regional production systems** (Scott, Storper) • **Industrial districts** (Piore & Sable)
Knowledge accumulation	• **Creative regions** (Andersson) • **Entrepreneurial regions** (Johannisson)	• **Learning regions** (Saxenian, Morgan) • **Innovative milieux** (Aydalot, Maillat) • **Industry clusters** (Porter)

Figure 6.1 Forces of agglomeration and spatial clustering

Both agglomeration in general and spatial clustering of related firms and industries have, as indicated above, traditionally been seen as being driven by efficiency considerations. In particular, explanatory models are based on economies of scale in transportation and transaction costs. Such approaches are situated in the upper part of the matrix. A more recent line of research has built models around the notion of flexible production systems in a more fast-moving world (Piore and Sabel, 1984), but we would argue that the corresponding agglomeration model - that of the industrial district - is still located in the upper right hand cell.

At present, however, there seems to be a tendency to assume that the benefits of agglomeration are more subtle and of a social rather than purely economic nature, and that the key to understanding both agglomeration in general and spatial clustering of related industries in particular lies in the superior ability of such spatial configurations to enhance learning, creativity and innovations, defined in a broad sense. This would mean a general shift of attention towards the lower part of the matrix. Spatial clustering is argued to reinforce mutual commitments and moderate inclinations toward opportunism, where trust may be viewed as a localised non-tradeable business input. Furthermore, knowledge spill-overs are accelerated in spatial clusters of related industries, and tacit knowledge tends to become embedded in local milieux (Dosi, 1988; Lundvall, 1988). In the following section we go on to explore some of the dynamics taking place in the lower right-hand cell of the matrix, i.e. in spatial clusters of related firms, embedded in local milieux characterised by particularly strong abilities to accumulate knowledge.

Local accumulation of knowledge

While agglomeration theory has picked up on the notion of knowledge accumulation in explaining the emergence and sustainability of spatial clusters of related firms and industries, innovation literature provides a complementary illustration of the very same process. Here, we propose a model of local accumulation of knowledge in a world of Schumpeterian competition (Schumpeter, 1942), arguing that the accumulation of knowledge and sustained competitiveness can be understood as a function of three interrelated processes. First, there is the localised nature of innovation processes and the role of local milieux in fostering such processes. Second, is the process whereby knowledge tends to stick to the local milieu rather than being rapidly diffused and, third, is a process whereby new resources - in the form of people, capital, ideas, patents and so on - may be attracted into the local milieu.

Localised innovation processes and the role of the local milieu

The first element of the local accumulation of knowledge relates to the nature of the innovation process. Research on the innovation process has identified three interrelated characteristics which are particularly important for understanding spatial clustering of related firms (Freeman, 1991):

- the need for incremental reduction of technical and economic uncertainty;
- the need for continuous interaction between related firms; and
- the need for face-to-face contacts in the exchange and creation of new knowledge.

The first characteristic derives from the fact that innovative processes are fundamentally uncertain in terms of technical feasibility and market acceptance (Freeman, 1982; Pearson, 1991). There is evidence that only one out of ten research projects turns out a commercial success, and that many patented inventions never find any direct commercial applications (Pavitt, 1991; Schmookler, 1966; Basberg, 1987). Although the level of uncertainty varies with the type of invention the technical aspects are commonly worked out by means of trial and error testing and modification. Incrementalism and trial and error problem solving in turn leads to a need for continuous interaction, both in informal networks and formal cooperative agreements, with related firms and other important public and private organisations such as universities and research centres.

The second feature of the innovation process is that ideas frequently originate outside the firm that carries out the actual development or manufacturing work (Pavitt, 1984). In fact, only a small proportion of all innovations has been found to be directed towards use within the inventing firm(s) and for improving internal processes (Scherer, 1984). The importance of customers as sources of innovation has been identified in several studies (Håkansson, 1989; Laage-Hellman, 1989), while others have added evidence that the development of functionally useful innovations is sometimes dominated by the suppliers (von Hippel, 1988). In yet other cases, several firms might be involved in joint development work by which each of the participants supplies a limited component of the resulting innovation. This makes the innovation process highly interactive - between firms and the basic science infrastructure, between producers and users at the inter-firm level, and between firms and their wider institutional milieu (Lundvall, 1988; Morgan, 1995).

Technological influences take on many forms, ranging from the one-time transfer of information to more extensive interaction between individual firms. However, there are reasons to believe that repeated interaction and lasting arrangements for the exchange of knowledge between firms, or technical learning, is very common in the innovation process. The importance of long-term producer and buyer relationships for the exchange and creation of knowledge has been noted by several authors (Lundvall, 1988, 1993; Håkansson and Eriksson, 1993; Hallén, Johanson and Seyed-Mohamed, 1993). This exchange frequently involves sensitive information, which might cause damage if used opportunistically by the firms involved and, therefore, requires a high level of trust between the parties. Similar linkages between the scientific community and firms engaged in technological improvements have also been illustrated (Freeman, 1982)[5].

The third characteristic of innovation is the employment of informal mechanisms for knowledge exchange. In spite of increasingly sophisticated means of communication, the need for personal, face-to-face contacts in the exchange of information has not disappeared (Törnqvist, 1970; Fredriksson and Lindmark, 1979; Nohria and Eccles, 1992). Also, personal contacts have been identified as important sources of technological information and improvements in the innovation process (Leonard-Barton, 1982; de Meyer, 1991; 1992). Moreover, traditional geographic wisdom suggests that there is 'friction of distance', implying that the probability of interpersonal communication through face-to-face contacts declines with increasing distance between individuals (Hägerstrand, 1967; Pred, 1977).

Face-to-face contacts appear to be of particular value for exchanging tacit knowledge, or when the exchange of knowledge involves direct observation of products or production processes that are in use. This type of knowledge typically does not reside in blueprints and formulae, but is based on personal skills and operational procedures which do not lend themselves to be presented and defined in either language or writing (Polanyi, 1962; Winter, 1987). Some studies indicate that informal and oral information sources provide most key communications about the market opportunities and technological possibilities that lead to innovation. According to Utterback (1974), unanticipated, or unplanned personal encounters often turn out to be most valuable. It is in this context that the geographically concentrated industrial configuration has substantial advantage over a dispersed configuration (Enright, 1994).

In summary, the very nature of the innovation process tends to make technological activity locally confined, and suggests that recent globalisation forces have not altered - and presumably cannot alter in the near future - this process in any fundamental way. In particular, the costs and time associated with repeated exchange of knowledge and information in the development work will be lowered if taking place in the local context (Zander and Sölvell, 1995). These aspects are important determinants of success in Schumpeterian competition, as reduced costs and shortened development times increase the size and length of the temporary monopolies that firms can achieve.

System configuration and knowledge generation

So far, we have argued that learning in many ways is to be regarded as a collective process, following from repeated interaction within and across firms. We now turn more explicitly to the question of the relation between the spatial configuration of an industrial system and its ability to generate new knowledge. The argument to be advanced is that, when a cluster of related activities develops in a particular place, the local society will tend to adjust in ways that might add further dynamism to the cluster itself, through the emergence of a local milieu.

We are deliberately vague on how to define concepts such as local or regional in this analysis. A spatial cluster is seen as a set of inter-linked firms and/or activities that exist in the same local or regional milieu, defined as to encompass economic, social, cultural and institutional factors. A local milieu could thus be a nation - this is indeed the normal use of the concept in the international business literature - but it could also be an urban region, or any functionally defined sub-national entity. The important issue in this context is not the precise delimitation of the local or regional milieu in geographical terms. It is sufficient for our purpose to make clear that it is a segment of territory characterised by a certain coherence based on common behavioural practices linked to its local institutions and culture (including 'technical culture'), industrial structure and corporate organisation (Saxenian, 1994).

Some basic characteristics of such milieux may be identified.[6] Thus, there are several actors (firms, institutions) that are relatively autonomous in terms of decision making and strategy formulation, and the interaction between these actors contains an element of both cooperation and rivalry. Furthermore, the milieu is characterised by a specific set of material (firms, infrastructure), immaterial (knowledge, know-how), and institutional (authorities, legal framework) elements. These elements make up a complex web of relations that tie firms, customers, research institutions, the school system, and local authorities to each other. The interaction between economic, socio-cultural, political and institutional actors in a given place may trigger learning dynamics and enhance the ability of actors to modify their behaviour and find new solutions as their competitive environment changes. This view of local milieux is often associated with the 'innovative milieu approach' (Aydalot, 1986; Maillat, 1995).

In such a perspective, a place or a region is not seen merely as a 'container' in which attractive location factors may happen to exist or not, but rather as a milieu for collective learning through intense interaction between a broadly composed set of actors. The milieu is both a result of and a precondition for learning - an asset that is constantly being produced and reproduced - rather than a passive surface (Coffey and Bailly, 1996). It is an environment within which physical capital and human capital is created and accumulated over time, which translates into sustainable competitiveness among incumbent firms.

In the local milieu, the fluidity of knowledge will be improved by the development of a common code of communication and interaction, particularly when knowledge is difficult or costly to codify. To a large extent, a common location will offer language and cultural similarities which improve the ease of communication. The local milieu thus offers an environment for the evolution of a common language, social bonds, norms, values and institutions, i.e. a 'social capital' (Putnam, 1993) which adds to the process of accumulated learning. Within the local milieu, these institutional arrangements become increasingly specialised and unique, adding to the fluidity of knowledge exchange.

Barriers to the diffusion of knowledge

The second element of local knowledge accumulation relates to inertia and various barriers to the diffusion of knowledge. Inertia has, as we have already seen, always had a central place in the geographic literature. However, it is also central to how easily knowledge which is embedded in one local milieu can be imitated by outside actors. If diffusion is indeed rapid and can be accomplished at low cost, much of our earlier argument would fall flat. If, on the other hand, it is possible to argue that diffusion in effect is sluggish, carries large costs and involves long lead times, the model of localised innovation and sustainability holds.

In order to understand the speed of the diffusion of knowledge, it is necessary to recognise the different degrees of knowledge mobility embedded in physical, human and social capital. With reduced trade barriers in the post-war period, knowledge embedded in standard materials, components, products and machinery move relatively easily and at low cost. However, while modern flexible machinery is being adopted at a high rate in most industrialised countries, empirical studies have indicated that firms may experience considerable difficulty in implementing even this type of technology effectively. These problems are aggravated when the buying firms are distant - physically and culturally - from the industrial milieux where such new process technologies are developed and produced (Gertler, 1995). Furthermore, incentives to export cutting-edge components and products may be relatively weak as they are fundamental to temporary monopolies and competitive advantage among the firms that developed them.

While some knowledge is embedded in materials, components, products and machinery, other knowledge is embedded in human capital, part of which is tacit. As a result of improved air transportation, an increasing proportion of skilled human capital, such as top management and experts, has become internationally mobile. However, an important part of human capital is embedded in a multitude of inter-firm relationships and therefore cannot be taken out of context without losing much of its value. These relationships include both formal and informal networks in the local milieu and have often been established and deepened over long periods of time. Furthermore, the large groups of middle and lower level managers and workers, who make up an important part of the formal and informal relationships between firms, are typically much less mobile than top management and experts.

The formal and informal networks between people in a common location, which have often been developed through long term interaction, and the resulting evolution of local institutions, form part of the social capital surrounding innovation processes. Whereas some knowledge embedded in physical and human capital to an increasing extent travels the world through trade, investment, travelling and migration, knowledge embedded in social capital does not, as it involves a large number of actors within a local milieu and is historically bound to local circumstances, involving unique bonds and accumulated routines (see Figure 6.2).

Thus, the diffusion of knowledge within the local milieu is rapid, whereas it is typically slow between one local milieu and another. With reference to Hägerstrand's (1967) model of diffusion, 'expansion diffusion' would be smooth and involve learning between firms in a local cluster, whereas hierarchical or sideways diffusion across milieux would be retarded. As a result of this, milieu embedded knowledge will lead to sustainable advantage as it is scarce, non-imitable, and non-substitutable

(Enright, 1994; Maskell and Malmberg, 1995). Under such conditions, the local milieu earns Ricardian rents on its resources, an effect which also spills over into the firms which are active within its confines.

Type of knowledge \ Mobility	High	Low
Knowledge embedded in physical capital	Materials Standard components and products Standard machinery Data in databases Blueprints CAD/CAM systems Published research	Cutting-edge components, products Cutting-edge machinery
Knowledge embedded in human capital	Top management Experts	Middle management Informal and formal networks
Knowledge embedded in social capital	Legal frameworks	History-bound routines Business practices Unique institutions Many linked actors

Figure 6.2 The Mobility of Different Types of Knowledge

Attracting knowledge to the local milieu

A third element in the model of the local accumulation of knowledge involves the attraction of outside resources. As the local milieu evolves it might attract new people, technology and firms from the outside, setting off an even stronger cumulative process. To use the Hollywood film cluster as an example, we might think of this as the 'Greta Garbo-effect'. The Swedish-born actress was attracted by the creativity of the Hollywood milieu. By moving there, she was able to develop her acting skills and become a leading world movie star. At the same time she added to the further artistic development of the milieu and created returns which could be reinvested in new projects.

The inflow of knowledge is driven both by actors from the outside being lured or attracted into the local milieu, and by local actors who try to tap into outside knowledge (see discussion below on the role of transnational corporations). Knowledge inflow is limited by the extent to which resources are mobile, which of course is the mirror argument of why diffusion out of a local milieu is sluggish. As we indicated above some physical and human capital can easily be sourced world-wide, but it is more difficult to tap into cutting-edge components, products and machinery or the knowledge which is embedded and exchanged within formal and informal networks in other milieux. Tapping becomes particularly limited in the case of social capital, which is extremely difficult to transfer and replicate from one milieu to another.

A conceptual model of knowledge creation and accumulation

In summary, the model we have suggested to explain the local accumulation of knowledge and sustainability of competitiveness involves three elements. The first has to do with the nature of the innovation process itself. Together, the trial and error process of problem solving, the repeated interaction between related firms, and the need to exchange knowledge through face-to-face contacts, provide an interrelated set of factors which favours locally confined innovation processes. The second element has to do with barriers to diffusion of locally embedded knowledge. In essence, the ability to gain access to informal and formal networks for knowledge exchange and to accumulated social capital is by and large reserved to the insiders in a milieu, and this is the ultimate barrier to the diffusion of knowledge. The third element, finally, is related to the attraction of outside resources enhancing the process of knowledge accumulation within the local milieu, involving both resources brought in by outsiders and initiatives taken by incumbents to tap resources from the outside (see Figure 6.3).

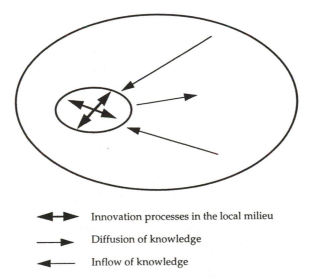

```
◀━━▶   Innovation processes in the local milieu

━━━▶   Diffusion of knowledge

◀━━━   Inflow of knowledge
```

Figure 6.3 Three processes involved in the local accumulation of knowledge

131

Transnational corporations and local milieux

Explicitly or implicitly, most models of industrial districts and innovative milieux are preoccupied with small or medium sized firms which are strongly place-based. The concept of industrial districts is partly defined by vertical disintegration, intense subcontracting relations, and extensive new firm formation through spin-offs (Asheim, 1992). The innovative milieu model also emphasises the inclusion of a large number of relatively autonomous firms.

International business scholars, in contrast, have studied the emergence of large transnational corporations (TNCs) during the past century. TNCs have today come to dominate many industrial sectors, and we can also discern a pattern of small and medium sized firms becoming transnational at a very early stage. The modern TNC is typically built around a set of global activities, including centres for corporate and business unit headquarters, purchasing, marketing and sales, manufacturing, and research and development. If we are right in our analysis of the localised accumulation of knowledge, then the question naturally arises as to how this would be affected by the presence and operations of TNCs. Can the well-established TNC build competitive advantage through the international integration of innovative activities in several local milieux, thereby reducing the strength and sustainability of these milieux?

Leading TNCs have begun to establish major operations outside their home countries as a result of long term investment abroad, but more often through foreign mergers, acquisitions and strategic alliances (Dunning, 1993). Sometimes, divisional headquarters and all development activities for certain business areas are concentrated in local milieux outside the home country (Dunning, 1994; Zander, 1994; Cantwell, 1995). These diversified TNCs have thus created something of a 'multi home-base' structure (Forsgren, Holm and Johanson, 1991; Sölvell and Zander, 1995), involving several distinct bases for innovation which are often referred to as centres of competence or centres of excellence.

In local milieux where the TNC controls full-fledged operations, the corporation can be characterised as an insider just like any other local firm. It is linked to other firms in both formal and informal networks, and typically maintains close linkages with local research and education facilities, governmental bodies and other important actors. These linkages provide the channels for the rapid dissemination of knowledge and information, and provide a basis for cooperation leading to a continuous stream of small and large improvements. Key facilitators of knowledge flows would include personal relationships that arise from schooling and military service, the mobility of employees between competing firms, norms of behaviour supporting continuity and long term relationships, or quasi-family ties between firms and interlocking directorates. Sometimes, TNCs have built up insider positions through long term investment, but more often TNCs become insiders by acquiring local firms with full fledged operations and established local networks.[7]

Along these lines, some authors have argued that TNCs which have built insider positions in several local milieux are now becoming engaged in the integration of innovative activities across their geographically dispersed units (Prahalad and Doz, 1987; Bartlett and Ghoshal, 1990; Hedlund and Rolander, 1990). Implicitly, these authors stress the development of an 'organisational capital' within the TNC organisation, by which a common set of norms, values and routines makes it possible to overcome differences in social capital across local milieux (Hedlund, 1986). In well-established TNCs, the geographically dispersed network of subsidiaries becomes a means for rapid knowledge exchange, leading to the development of unique advantages from the integration of the global corporate system.

This line of research, we believe, has underestimated the increasing costs and lengthened development times associated with innovation across geographical distances, which will be a significant disadvantage in global competition. Also, several authors have emphasised the difficulties involved in creating a set of common norms, values and working routines in the TNC that are necessary for cross-border innovation to take place (Kilduff, 1992; Håkanson, 1995; Holm, Johanson and Thilenius, 1995). If, indeed, the TNC faces inefficiencies in internationally co-ordinated innovation processes, it would retain its character as a local innovator (for global commercialisation), thereby adding to the strength of the local milieux in which it is active. In the case of the diversified TNC, we might expect that each business segment would concentrate its core resources to the most dynamic local milieu.

Concluding remarks

A review of theories of agglomeration leads us to two conclusions. First, it is important to distinguish between forces that lead to patterns of uneven regional development in general (i.e. the emergence of industrial centres and peripheries at a global and national scale) and forces that tend to make related firms and industries cluster spatially. Second, when it comes to understanding the forces behind and the effects of spatial clustering, it is important to focus upon the knowledge accumulation effects of spatial clustering rather than exclusively upon potential benefits in terms of (short term) transaction efficiency and mere flexibility.

Turning to the issue of knowledge accumulation within spatial clusters, we have argued that it is composed of three equally important processes. The first process has to do with the ability to generate new knowledge through various kinds of interaction in a local milieu. The aspects of the innovation process of vital importance here are its intrinsically uncertain and interactive character and the important role played by the exchange of tacit knowledge - best communicated by means of face-to-face contacts. The second process has to do with barriers to the diffusion of knowledge from the local milieu, and the third relates to the ability to attract new knowledge to a milieu from outside to enhance local dynamism.

Finally, we argue that TNCs, rather than being in opposition to the notion of local knowledge accumulation, to a large degree follow a similar 'learning logic'. They too are dependent upon strong local milieux - or home bases - in the knowledge accumulation necessary for their long term competitiveness. Thus, their increasing importance in the world economy does not diminish the importance of the arguments advanced here. To the contrary, it increases their relevance.

Acknowledgements

This paper draws upon research carried out as part of the authors' involvement in a Swedish research programme on *Regions in International Competition*, funded by the Swedish Institute for Regional Research (SIR), and the Swedish Council for Research in the Humanities and Social Sciences (HSFR).

Notes

1.   The content of this paper is overlapping with a previously published paper (Malmberg, Sölvell and Zander 1996). For a similar account, see also Chandler, Hagström and Sölvell (1997).

2.   See Malmberg (1996) for a review of traditional and recent contributions to this body of theory.

3.   This aspect of the agglomeration phenomenon has been brought to the fore in recent years, but it has also a longer history in academic research, see e.g. Pred (1977).

4.   Three types of externalities that lead to the localisation of particular industries may be derived from Marshall. First, a pooled market develops for workers with industry specific skills, which can be assumed to lead to lower probability both of unemployment among workers and of labour shortage among firms. Second, the production of non-tradeable inputs is supported, and third, informational spill overs occur, which gives clustered firms a better production function (cf. Krugman, 1991a, 1991b).

5.   Long term and stable relationships between firms are an important determinant of technological path-dependency, which has been identified both at the firm and national levels (Dosi, 1982; Pavitt, Robson and Townsend, 1989; Pavitt, 1988; Cantwell, 1991; Archibugi and Pianta, 1992).

6.   We do not intend here to give a full review of the by now fairly extensive literature on innovative milieux (for broader accounts, see e.g. Camagni (1991), Conti (1993) or Maillat (1995)). Our aim in the remainder of this section is simply to establish the fact that the localised learning processes in focus are embedded in a wider societal context - in local milieux - that may under certain conditions enhance the learning process.

7.   The argument here is, as indicated above, focused on the innovative activities of the TNC. Much research on the local and regional impacts of TNC investments have been preoccupied with less fully-fledged corporate operations, such as branch plants involved in routinised manufacturing, which may admittedly be less embedded in the local milieu. A broad discussion of the local embeddedness of TNCs is provided in Dicken et al. (1994).

References

Andersson, Å.E. (1985), 'Creativity and regional development', *Papers of theRegional Science Association*, vol. 56, pp. 5-20.

Amin, A. and Malmberg, A. (1992), 'Competing structural and institutional influences on the geography of production in Europe', *Environment and Planning A*, vol. 24, pp. 401-416.

Amin, A. and Thrift, N. (1995), 'Territoriality in the global political economy', *Nordisk Samhällsgeografisk Tidskrift*, vol. 20, pp. 3-16.

Appold, S.J. (1995), 'Agglomeration, interorganizational networks, and competitive performance in the U.S. metalworking sector', *Economic Geography*, vol. 71, pp. 27-54.

Archibugi, D. and Pianta, M. (1992), 'Specialization and size of technological activities in industrial countries: the analysis of patent data', *Research Policy*, vol. 21, pp. 79-93.

Asheim, B.T. (1992), 'Flexible specialization, industrial districts and small firms: a critical reappraisal', in Erneste, H. and Maier, V. (eds), *Regional Development and Contemporary Industrial Response. Extending Flexible Specialization*, Belhaven Press, London.

Aydalot, P. (1986), *Milieux Innovateurs en Europe*, GREMI, Paris.

Bartlett, C.A. and Ghoshal, S. (1990), 'Managing innovation in the transnational corporation', in Bartlett, C.A., Doz, Y. and Hedlund, G. (eds), *Managing the Global Firm*, Routledge, London.

Basberg, B.L. (1987), 'Patents and the measurement of technological change: a survey of the literature', *Research Policy*, vol. 16, pp. 131-141.

Beccattini, G. and Rullani, E. (199X), 'Local systems and global connections: the role of knowledge', in Cossentino, F., Pyke, F. and Sengenberger, W. (eds), *Local and Regional Response to Global Pressure: The Case of Italy and its Industrial Districts*, Research Series 103, International Institute for Labour Studies, Geneva.

Camagni, R. (ed.) (1991), *Innovation Networks: Spatial Perspectives*, Belhaven Press, London.

Cantwell, J. (1991), 'Historical trends in international patterns of technological innovation', in Foreman-Peck, J. (ed.), *New Perspectives on the Late Victorian Economy. Essays in Quantitative Economic History 1860-1914*, Cambridge University Press, Cambridge.

Cantwell, J. (1995), 'The globalization of technology: what remains of the product cycle model?', *Cambridge Journal of Economics*, vol. 19, pp. 155-174.

Chandler Jr, A., Hagström, P. and Sölvell, Ö. (1997), *The Dynamic Firm*, Oxford University Press, Oxford. (forthcoming)

Coffey, J. and Bailly, A. (1996), 'Flexible economy and services: new patterns of location of economic activity', in Bailly, A. and Lever, W.F. (eds), *The Spatial Impacts of Economic Changes in Europe*, Avebury, Aldershot.

Conti, S. (1993), 'The network perspective in industrial geography: towards a model', *Geografiska Annaler Series B*, vol. 75B, pp. 115-130.

Dicken, P. (1992), *Global Shift. The Internationalization of Economic Activity*, Second edition, Paul Chapman Publishing, London.

Dicken, P., Forsgren, M. and Malmberg, A. (1994), 'The local embeddedness of transnational corporations', in Amin, A. and Thrift, N. (eds), *Globalization, Institutions, and Regional Development in Europe*, Oxford University Press, Oxford.

Dosi, G. (1982), 'Technological paradigms and technological trajectories', *Research Policy*, vol. 11, pp. 147-162.

Dosi, G. (1988), 'Sources, procedures and microeconomic effects of innovation', *Journal of Economic Literature*, vol. XXVI, pp. 1120-1171.

Dunning, J.H. (1993), *Multinational Enterprise and the Global Economy*, Addison-Wesley, Wokingham.

Dunning, J.H. (1994), 'Multinational enterprises and the globalization of innovatory capacity', *Research Policy* 23, pp. 67-88.

Enright, M.J. (1994), 'Regional clusters and firm strategy', paper presented at the Prince Bertil Symposium on The Dynamic Firm: The Role of Regions, Technology, Strategy and Organization, Stockholm, 12-15 June.

Estall, R.C. and Buchanan, R.O. (1961), *Industrial Activity and Economic Geography*, Hutchinson, London.

Forsgren, M., Holm, U. and Johanson, J. (1991), 'Internationlisering av andra graden', in Andersson, R. et al. (eds), *Internationalisering, företagen och det lokala samhället*. SNS Förlag, Stockholm.

Fredericksson, C. and Lindmark, L. (1979), 'From firms to systems of firms: a study of interregional dependence in a dynamic society', in Hamilton, F.E.I. and Linge, G.J.R. (eds), *Spatial Analysis, Industry and the Industrial Environment - Progress in Research and Applications: Vol. 1 - Industrial Systems*, Wiley, Chichester.

Freeman, C. (1982), *The Economics of Industrial Innovation*, Second edition, Frances Pinter Publishers, London.

Freeman, C. (1991), 'Networks of innovators: a synthesis of research issues', *Research Policy*, vol. 20, pp. 499-514.

Gertler, M. S. (1995), '"Being there": Proximity, organization, and culture in the development and adoption of advanced manufacturing technologies', *Economic Geography*, vol. 71, pp. 1-26.

Hallén, L., Johanson, J. and Seyed-Mohamed, N. (1993), 'Dyadic business relationships and customer technologies', *Journal of Business-to-Business Marketing*, vol. 1, pp. 63-90.

Hedlund, G. (1986), 'The hypermodern MNC - a heterarchy?', *Human Resource Management*, vol. 25, pp. 9-35.

Hedlund, G. and Rolander, D. (1990), 'Action in heterarchies: new approaches to managing the MNC', in Bartlett, C.A., Doz, Y. and Hedlund, G. (eds), *Managing the Global Firm*, Routledge, London.

von Hippel, E. (1988): *The Sources of Innovation*, Oxford University Press, Oxford.

Hirschman, A.O. (1958), *The Strategy of Economy Developmen,.* Yale University Press, New Haven.

Holm, U., Johanson, J. and Thelenius, P. (1995), 'Headquarters' knowledge of subsidiary network contexts in the multinational corporation', *International Studies of Management and Organization*, vol. 25, pp. 97-119.

Hoover, E.M. (1948), *The Location of Economic Activity,* McGraw-Hill, New York.

Håkansson, H. (1989), *Corporate Technological Behaviour - Co-operation and Networks*, Routledge, London.

Håkansson, H. and Eriksson, A.-K. (1993), 'Getting innovations out of supplier networks', *Journal of Business-to-Business Marketing*, vol. 1, pp. 3-34.

Håkanson, L. (1995), 'Learning through acquisitions: management and integration of foreign R&D laboratories', *International Studies of Management and Organization*, vol. 25, pp. 121-157.

Hägerstrand, T. (1967), *Innovation Diffusion as a Spatial Process*, University of Chicago Press, Chicago.

Johannisson, B. (1987), 'Toward a theory of local entrepreneurship', in Wykman, R. G., Merredith, L. N. and Bush, G. R. (eds), *The Spirit of Entrepreneurship*, Simon Fraser University, Vancouver, BC.

Kilduff, M. (1992), 'Performance and interaction routines in multinational corporations', *Journal of International Business Studies*, vol. 23, pp. 133-145.

Kogut, B. (ed.) (1993), *Country Competitiveness - Technology and Organizing of Work*, Oxford University Press, Oxford.

Krugman, P. (1991a), *Geography and Trade*, MIT Press, Cambridge, MA.

Krugman, P. (1991b): 'Increasing returns and economic geography', *Journal of Political Economy*, vol. 99, pp. 483-499.

Laage-Hellman, J. (1989), *Technological Development in Industrial Networks*, Acta Universitatis Upsaliensis, No. 16, Faculty of Social Sciences, Uppsala University, Uppsala.

Leonard-Barton, D. (1982), *Swedish Entrepreneurs in Manufacturing and Their Sources of Information*, Center for Policy Applications, MIT, Boston.

Lloyd, P.E. and Dicken, P. (1977), *Location in Space. A Theoretical Approach to Economic Geograph,.* Second edition, Harper and Row, London.

Lundvall, B.-Å. (1988), 'Innovation as an interactive process: from user-producer interaction to the national system of innovation', in Dosi, G. et al. (eds), *Technical Change and Economic Theory*, Pinter, London.

Lundvall, B.-Å. (ed.) (1992), *National Systems of Innovation: Towards a Theory of Innovation and Interactive Learning*, Pinter, London.

Lundvall, B.-Å. (1993), 'Explaining interfirm cooperation and innovation: limits of the transaction-cost approach', in Grabher, G. (ed.), *The Embedded Firm - On the Socioeconomics of Industrial Networks*. Routledge, London.

Maillat, D. (1995), 'Territorial dynamic, innovative milieus and regional policy', *Entrepreneurship and Regional Development*, vol. 7, pp. 157-165.

Malecki, E. J. (1991), *Technology and Economic Development: The Dynamics of Local, Regional and National Change*, Longman, New York.

Malmberg, A. (1996), 'Industrial geography: agglomeration and local milieu', *Progress in Human Geography*, vol. 20 (forthcoming).

Malmberg, A. and Maskell, P. (1996), 'Proximity, institutions and learning - towards an explanation of regional specialization and industry agglomeration', Paper presented at the EMOT workshop, Durham University, June.

Malmberg, A., Sölvell, Ö. and Zander, I. (1996), 'Spatial clustering, local accumulation of knowledge and firm competitiveness', *Geografiska Annaler Series B*, vol. 78B.

Marshall, A. (1890/1916), *Principles of Economics. An Introductory Volume*, Seventh edition, Macmillan, London.

Maskell, P. and Malmberg, A. (1995), 'Localised learning and industrial competitiveness', *BRIE Working Paper* 80, Berkeley Roundtable on the International Economy, University of California, Berkeley.

McCann, P. (1995), 'Rethinking the economics of location and agglomeration', *Urban Studies*, vol. 32, pp. 563-577.

de Meyer, A. (1991), 'Tech talk: how managers are stimulating global R&D communication', *Sloan Management Review*, Spring, pp. 49-58.

de Meyer, A. (1992), 'Management of international R&D operations', in Granstand, O., Håkanson, L. and Sjölander, S. (eds), *Technology Management and International Business: Internationalization of R&D and Technology*, John Wiley and Sons, Chichester.

Morgan, K. (1995), 'Institutions, innovation and regional renewal. The development agency as animateur', paper presented at the Regional Studies Conference on Regional Futures, Gothenburg, 6-9 May.

Mydal, G. (1957), *Economic Theory and the Underdeveloped Regions*, Ducksworth, London.

Nohia, N. and Eccles, R.G. (1992), 'Face-to-face: making network organizations work', in Nohria, N. and Eccles, R.G. (eds), *Networks and Organizations: Structure, Form, and Action*, Harvard Business School Press, Boston, MA.

Patchell, J. (1993), 'From production systems to learning systems: lessons from Japan', *Environment and Planning A*, vol. 25, pp. 797-815.

Patel, P. (1995), 'Localised production of technology for global markets', *Cambridge Journal of Economics*, vol. 19, pp. 141-153.

Pavitt, K. (1984), 'Sectoral patterns of technical change: towards a taxonomy and a theory', *Research Policy*, vol. 13, pp. 343-373.

Pavitt, K. (1988), 'International patterns of technological accumulation', in Hood, N. and Vahlne, J.-E. (eds), *Strategies in Global Competition*, Croom Helm, London.

Pavitt, K. (1991), 'Key characteristics of the large innovating firm', *British Journal of Management*, vol. 2, pp. 41-50.

Pavitt, K., Robson, M. and Townsend, J. (1989), 'Technological accumulation, diversification and organisation in UK companies 1945-1983', *Management*

139

Science, vol. 35, pp. 81-99.

Pearson, A. W. (1991), 'Managing innovation: an uncertainty reduction process', in Henry, J. and Walker, D. (eds), *Managing Innovation*, Sage, London.

Piore, M. and Sabel, C. (1984), *The Second Industrial Divide*, Basic Books, New York.

Polanyi, K. (1962), *Personal Knowledge: Towards a Post-Critical Philosophy*, Chicago University Press, Chicago.

Porter, M.E. (1990), *The Competitive Advantage of Nations*, Macmillan, London and Basingstoke.

Porter, M. E. (1994), 'The role of location in competition', *Journal of the Economics of Business*, vol. 1, pp. 35-39.

Prahalad, C.K. and Doz, Y. (1987), *The Multinational Mission - Balancing Local Demands and Global Vision*, The Free Press, New York.

Pred, A. (1977), *City Systems in Advanced Economies. Past Growth, Present Processes and Future Development Options*, Hutchinson, London.

Putnam, R. D. (with Leonardi, R. and Nanetti, R. Y.) (1993), *Making Democracy Work. Civic Traditions in Modern Italy*, Princeton University Press, Princeton.

Ridderstråle, J. (1996), *Global Innovation - Managing International Innovation Pprojects in ABB and Electrolux,* Doctoral dissertation, Institute of International Business, Stockholm School of Economics, Stockholm.

Saxenian, A. (1994), *Regional Advantage. Culture and Competition in Silicon Valley and Route 128*, Harvard University Press, Cambridge, MA..

Scherer, F.M. (1984), *Innovation and Growth - Schumpeterian Perspectives*, MIT Press, Cambridge, MA.

Schmookler, J. (1966), *Inventions and Economic Growth*, Harvard University Press, Cambridge, MA.

Schumpeter, J. A. (1942), *Capitalism, Socialism and Democracy*, Harper and Row, New York.

Scott, A.J. (1983), 'Industrial organisation and the logic of intra-metropolitan location - 1. Theoretical considerations', *Economic Geography*, vol. 59, pp. 233-250.

Scott, A.J. (1988), *New Industrial Spaces: Flexible Production Organisation and Regional Development in North America and Western Europe*, Pion, London.

Scott, A.J. (1995), 'The geographic foundations of industrial performance', *Competition and Change, The Journal of Global Business and Political Economy*, vol. 1, pp. 51-66.

Storper, M. and Walker, R. (1989), *The Capitalist Imperative. Territory, Technology, and Industrial Growth*, Basil Blackwell, New York.

Storper, M. (1995), 'The resurgence of regional economies, ten years later: the region as a nexus of untraded interdependencies', *European Urban and Regional Studies*, vol. 2, pp. 91-221.

Sölvell, Ö., Zander, I. and Porter, M.E. (1991), *Advantage Sweden*, Norstedts,

Stockholm.

Sölvell, Ö. and Zander, I. (1995), 'Organization of the dynamic multinational enterprise - the home-based and the heterarchical MNE', *International Studies of Management and Organization*, vol. 25, pp. 17-38.

Törnqvist, G. (1970), *Contact Systems and Regional Development*, Lund Studies in Geography, Series B, No. 35, University of Lund, Lund.

Ullman, E.L. (1958), 'Regional development and the geography of concentration', *Papers and Proceedings of the Regional Science Association*, vol. IV, pp. 179-198.

Utterback, J. (1974), 'Innovation in industry and the diffusion of technology', *Science*, vol. 183, pp. 658-662.

Weber, A. (1909/1929), *Theory of the Location of Industries*, The University of Chicago Press, Chicago.

Winter, S.G. (1987), 'Knowledge and competence as strategic assets', in Teece, D. (ed.), *The Competitive Challenge - Strategies for Industrial Innovation and Renewal*, Ballinger Publishing Company, Cambridge, MA.

Zander, I. (1994), *The Tortoise Evolution of the Multinational Corporation: Technological Activity in Swedish Multinational Firms 1890-1990*, published doctoral dissertation, Institute of International Business, Stockholm School of Economics, Stockholm.

Zander, I. and Sölvell, Ö. (1995), 'The local nature of the innovation process: implications for the multinational firm', paper presented at the EMOT Workshop, University of Reading, June.

141

7 'Learning regions' in a globalised world economy: towards a new competitive advantage of industrial districts?

Bjørn T. Asheim

Introduction

This chapter examines if we today, as a result of post-Fordist changes in the organisation and location of production, are experiencing a shift towards a new competitive advantage for regions of territorially agglomerated SMEs. This question is of strategic academic and political importance in the globalisation-localisation debate, since a positive answer would provide a solid basis for launching a policy for endogenous industrial and regional development.

According to Lipietz the global-local debate 'has taken a new turn: it is the debate between the approach in terms of the interregional (or international) division of labour and that in terms of endogenous development, of which a characteristic form would be the industrial district' (Lipietz, 1993, p. 8).

Some observers argue that globalisation tendencies are of particular importance in changing the world economy because 'the global strategies of individual MNEs [multinational enterprises] play an ever more crucial role in economic development, in shaping the world economy and in significantly influencing the direction, contact and outcome of public policy choices. MNEs have more power, not only in setting policy agendas and shaping outcomes in the international political economy, but also in a wider more pervasive sense, in influencing for good or bad the values of culture and society as well as the "life-chances" of the individual' (Sally, 1994, p. 163). Amin (1993) also argues that 'the principal direction of change within industry in Europe is not towards local agglomeration or the resurgence of regional economies. Instead, it is towards the elaboration of new forms of globalization the organization of industrial activity, combined with greater geographical centralization of both institutional and corporate power and control' (p. 279).

Leborgne and Lipietz (1992) argue that 'in the field of industrial organization, oppressive worldwide subcontracting will compete with dense, territorially based partnership' (p. 348). According to Cooke and Morgan (1994), the SMEs of industrial districts are increasingly being squeezed between 'more aggressive and more flexible strategies on the part of the large firms on the one hand and, on the other, by increased competition from low-wage countries' (p. 105). In this connection, however, it is important to remember that 'oligopolistic structures of production are quite compatible with increasing flexibility, and that flexibility is not a characteristic specific to small-scale, non-hierarchical, integrated production complexes' (Martinelli and Schoenberger, 1991, p. 117).

Generally, the combined effects of the globalised and deregulated world economy and the reduced power of nation states due to the transfer of authority to supranational organisations (e.g. the EU and the WTO) have resulted in an increased need for forms to establish organisational micro regulation in order to improve their ability to control the growing complexity and insecurity in the increasing competitive world economy.

This new micro regulation is achieved either through global or local networks as an independent, third form of governance which is an alternative to markets (of a globalised world economy) and hierarchies (of large corporations). In this connection it is obviously important to distinguish between global and local networks. Global networks are constituted by functionally integrated production systems dominated by large firms (TNCs), and are caused by a change from vertical integration of industrial production in traditional hierarchical large firms to vertical quasi-integration or disintegration. The typical local network is a territorially agglomerated, local production system established by SMEs (e.g. the industrial district and other innovative milieus as the ideal type.[1] Generally 'the network conceptualization implies ... that the boundaries of the firm need to be redefined' (Dicken, 1994, p. 105-106).

Thus, the aim of this chapter is to analyse the prospects of promoting endogenous development in regions dominated by SMEs in industrial districts, in the wider context of globalisation and the extension of the boundaries of TNCs. It is contended that the core issue is related to the collective learning capacity of SMEs in industrial districts, which is crucial to their innovativeness and flexibility (Johnson and Lundvall, 1991). The chapter will discuss factors enabling and constraining the formation of sufficient learning capacity to bring about the necessary structural change to secure the successful transformation of districts into 'learning regions' which could exploit 'the benefits of learning-based competitiveness' (Amin and Thrift, 1995, p. 11).

Globalisation, TNCs and the division of labour

Globalisation and the division of labour

A widening of the capitalist world economy could be described as a process of internationalisation. However, it is important to appreciate that 'the terms "international" and "global" are not used synonymously. Internationalisation refers

144

simply to the extension of activities across national boundaries; globalisation involves more than this and is quantatively different. It implies a degree of purposive functional integration among geographically dispersed activities' (Dicken, 1994, p. 106). This distinction could be further refined to imply *the new international division of labour* as a globalisation of *the production process* (or *production space*) changing geographical patterns of specialisation at the global scale and organised by and within TNCs. It also represents an important change in *the mode of integration* of the world economy. Before globalisation, the international economy was primarily based on nation states as the main organising units, producing an *inter*national economy. Today, TNCs as the main organising units of the world economy represent a change towards a functional integration at a supra-territorial level, which implies *globalised* economy. When using the concept 'the new international division of labour', we should refer to a globalisation of *the technical division of labour*, i.e. to an extension of (what was originally) the *internal* technical division of labour inside a production unit of a globalised production process within production systems organised by TNCs, and not to the diffusion of capitalist industrial production and trade with manufacturing products to countries in the Third World. Dicken et al (1994) maintain that 'a growing number of TNCs most notably ... the larger ones, have been developing globally integrated competitive strategies. ... In these global strategies the firms tend to unify the range of products in all national markets, to centralise the R&D process' (p. 11). This represents a qualitatively new phase in the development of the capitalist world economy.

The growing importance of the *technical division of labour* in global production systems governed by TNCs and, consequently, the geographical expansion of the arena of the economy in relation to the arena of politics of the nation state has reduced the possibilities of state intervention. The basis for state intervention in a Fordist/Keynesian mode of regulation has been the relative importance of the social division of labour in the economy, as the focus of regulation was the plan versus market dictotomy within the framework of nation states. As TNCs represent the dynamic force in the globalisation process, "governments are competing more in an industrial and technological race for the resources and capabilities of MNEs, in the process losing the power to pursue nationally isolated policies that do not take MNEs into account. ... This represents a shift from an agenda of regulation, of MNEs by government, to one of negotiation" (Sally 1994, p.178).

In general, the distinction between the social and the technical division of labour is misunderstood when tendencies towards vertical disintegration in the organisation of industrial production are considered to imply a broadening of the social division of labour; e.g. when Scott talks about "a localized network of producers bound together in a social division of labour" (Scott 1992, p.266). However, these changes are caused by an extension of the *technical* division of labour.[2]

This widespread misunderstanding seems to be caused by a misreading of Marx, focusing only on the *formal* argument of the independent ownership of subcontractors in a production system while completely ignoring the *substantial* argument of a *market* oriented production of commodities in order to be part of a social division of labour. The externalisation of specific tasks to a network of specialised contractors, which increases the level of vertical disintegration, must in my view be seen as an attempt to deepen what Marx called the despotism of the technical division of labour rather than as an extension of the anarchy in the social division of labour. [3]

TNCs and the question of embeddedness

Networking within local and global production systems produces new, planned forms of industrial organisation in contrast to the anarchic results of a market-based externalisation process. However, even if the organised forms of networking represent similarities, it is still crucial to recognise the differences in size and structure between global and local production systems. Thus, I disagree with Hirst and Zeitlin (1990) when they assume that the relatively autonomous productive units of big, decentralised corporations 'often resemble small, specialised firms or craft workshops' (p. 3). Such apparently *intra-firm* similarities cannot be sufficient to neutralise the qualitatively *inter-firm* differences between independent small firms located in industrial districts, cooperating with each other within formal and informal networks, on the one hand, and production units of large companies 'obtaining services ... from other divisions of the parent enterprise' (Hirst and Zeitlin, 1990, p.3) on the other. Correspondingly, it is also hard to comprehend how branch plants of large corporations could be 'both part of a multinational ... firm and an independent industrial district' (Sabel, 1989, p. 40).

This is basically a question whether TNCs are territorially embedded in regional and national economies or should be more appropriately conceptualised as operating in 'abstract economic spaces' (Perroux, 1970). Sally (1994) considers the TNC as 'the nodal point of and interface between two realms: that of internationalization in global structures, and that of embeddedness in the domestic structures of national/regional political economies' (p.162). He goes on to argue that 'it is in the home base that MNEs are most deeply embedded, where they have their headquarters operations and cores of value-adding activity, and where upstream activities in research, development, design and engineering tend to be concentrated. It is there that MNEs are most strongly linked in historically conditioned relationships with external actors such as local, regional and national governments, banks, trade unions, industry associations, suppliers and customers' (Sally, 1994, p.172). Also Dicken (1994) emphasises that 'TNCs are not placeless; all have an identifiable home base, a base that ensures that every TNC is essentially embedded within its domestic environment. Of course, the more extensive a firm's international operations, the more likely it will be to take on additional characteristics. ... But the point I am making is that TNCs are "produced" through a complex historical process of embedding, in which cognitive, cultural, social, political, and economic characteristics of the national home base play a dominant part. TNCs, therefore, are "bearers" of such characteristics, which then interact with the place-specific characteristics of the countries in which they operate to produce a particular outcome.

But the point is that the home-base characteristics invariable remain dominant' (Dicken, 1994, p. 117 and 118).

However, conceptually and substantially I disagree with this position on the embeddedness of TNCs. Both Sally's and Dicken's use of the concept of 'embeddedness' is not theoretically correct, and seems to be used in an everyday manner meaning simply 'located' or 'home base' as is illustrated when Sally (1994) reformulates Porters view of the home base of the MNE as 'the embeddedness of the MNE in its home base' (Sally, 1994, p. 181).[4] However, Granovetter (1985) talks about actors being 'embedded in networks of interpersonal relations' (Granovetter, 1985, p. 73) i.e. *social* integration, which means 'reciprocity between actors in context of co-presence' (Giddens, 1984, p. 28). Contrary to this, TNCs, as 'the most important economic institution of our times' (Sally, 1994, p. 169), are part of the economy, which is based on *system* integration, i.e. 'Reciprocity between actors or collectivities across extended time-space' (Giddens, 1984, p. 28), and represents the sphere for strategic, instrumental rationality.[5]

Industrial districts and endogenous technological development

Industrial districts and incremental innovations

Piore and Sabel (1984) highlighted permanent innovation as a vital characteristic of industrial districts, and a precondition for their continuous growth. According to Piore and Sable 'the fusion of productive activity, in the narrow sense, with the larger life of the community' represent 'the common solution' to the problems of 'the reconciliation of competition and cooperation' as well as of 'the regeneration of resources required by the collectivity but not produced by the individual units of which it is composed' (p. 275). The development of industrial districts confirms that such 'fusion' can solve the first problem, and also - conditioned by a supporting local organisational and institutional infrastructure - can increase the collective resources of a district. However, it is much more doubtful whether it, without further public intervention, has the potential to secure the permanent innovation and adoption of new technologies.

What Piore and Sabel emphasise is an understanding of industrial districts as a 'social and economic whole', where the success of the districts is as dependent on broader social and institutional aspects as on economic factors in a narrow sense (Pyke and Sengenberger, 1990). Bellandi emphasises that the economies of the districts originate from the thick local texture of interdependencies between the small firms and the local community (Bellandi,1989). Becattini maintains that 'the firms become rooted in the territory, and this result cannot be conceptualised independently of its historical development' (Becattini, 1990, p. 40). This 'Marshallian' view of the basic structures of industrial districts expresses the idea of 'embeddedness' as a key analytical concept in understanding the functioning of industrial districts (Granovetter, 1985). It is precisely the embeddedness in broader socio-cultural factors, originating in a pre-capitalist civil society, that is the material basis for Marshall's view of agglomeration economies as the specific *territorial* aspects of geographical agglomeration of economic activity (Asheim, 1992, 1994).

Thus, in contrast to traditional regional economics, Marshall attaches a more independent role to agglomeration economies. In traditional regional economics agglomeration economies are understood as agglomerated external economies, normally specified as localisation and urbanisation economies, i.e. it is used as a functional concept describing an intensification of the external economies of a production system by territorial agglomeration. For Marshall, external economies are obtained through the geographical concentration of groups of small firms belonging to the same industry (i.e. localisation economies), while in traditional regional economies the achievement of external economies of scale is not conditioned by a territorial agglomeration of industrial complexes. According to Perroux, it is possible to talk about growth poles in an 'abstract economic space' (Perroux, 1970). Thus, by defining agglomeration economies as social and territorial embedded properties of an area, Marshall abandons 'the pure logic of economic mechanisms and introduces a sociological approach in his analysis' (Dimou 1994, p. 27) . Also Harrison emphasises that this mode of theorising is fundamentally different from the one found in conventional regional economics or in any other neoclassical-based agglomeration theory (Harrison, 1991).

Marshall maintains that the two most important aspects of his understanding of agglomeration economies 'mutual knowledge and trust' and the 'industrial atmosphere', will together have a positive effect on the promotion of innovations and innovation diffusion among small firms within industrial districts. However, Marshall was also aware of the fact that agglomeration economies as such do not guarantee that product and process innovations will take place. Indeed, studies have shown that the 'industrial atmosphere' of industrial districts can support the adoption, adaptation and diffusion of innovations among SME's (Asheim, 1994). In the same way, the presence of trust can stimulate the introduction of new technology into industrial districts, since mutual trust - in addition to reducing transaction costs - seems to be crucial for the establishment of non-contractual inter-firm linkages.
However, the importance of territorial embedded agglomeration economies in promoting innovations concerns largely *incremental* innovations: 'Industrial districts can generate innovations by incremental steps, through a gradual improvement of the final product, of the process and of the overall production organization' (Bianchi and Giordani, 1993, p. 31). Garofoli also maintains that industrial districts have a larger capacity to deal with gradual innovations than with 'ruptures' (Garofoli, 1991a).

Thus, agglomeration economies can represent important basic conditions and stimuli to incremental innovations through informal 'learning-by-doing' and 'learning-by-using', primarily based on tacit knowledge (Asheim, 1994). This is conditioned by the productive balance between functional and territorial modes of integration (Asheim, 1994). It is precisely the combination of functionally integrated external economies and territorially integrated (Marshallian) agglomeration economies that have made the industrial districts so successful, but at the same time so vulnerable to changes in the international capitalist economy (Asheim, 1992).

As Bellandi suggests, such learning, based on practical knowledge (experience) of which specialised practice is a prerequisite, may have significant creative content (Bellandi 1994). Thus, as a result of what Bellandi calls 'decentralised industrial creativity' (DIC), the collective potential innovative capacity of small firms in industrial districts is not always inferior to that of large, research-based companies (Bellandi, 1994). Still the fact remains, however, that, in general, the individual results of DIC are incremental, even if 'their accumulation has possible major effects on economic performance' (Bellandi, 1994, p. 76).

Incremental innovations and competitive advantage

However, in an increasingly globalised world economy it is rather doubtful whether incremental innovations will be sufficient to secure the competitive advantage of SMEs in industrial districts. Crevoisier argues that the reliance on incremental innovations 'would mean that these areas will very quickly exhaust the technical paradigm on which they are founded' (Crevoisier, 1994, p. 259). In addition Bellandi underlines that 'consistency (between DIC and MID) does not mean necessity. A number of difficulties may arise which can constrain and even bring to a halt DIC within an industrial district' (Bellandi, 1994, p. 80-81).

The balance between functional and territorial integration promoting incremental innovations is threatened by stronger functional integration from an extension of the time-space distanciation of the production system of industrial districts through increased external ownership and control of local industry (Asheim, 1992). Such extensions can either be the result of large corporations buying up successful small enterprises in districts, and making them into branch plants or subsidiaries, thus transferring (final) decision-making to agencies external to the region; or as a result of the substitution for local subcontractors with firms outside the industrial districts in order to squeeze costs.

Concerning takeovers of SMEs by big companies 'it is important to observe how, by means of such operations, groups of enterprises and not only taking over individual firms, but through them can enter the web of relations between firms within the districts and integrate them into their corporate network' (Tolomelli, 1990, p.366).

Consequences of an externalisation of parts of the value chain are demonstrated by an example from the industrial district of Carpi in the Third Italy.

> In the 1980s the district reduced its vertical integration in the knitwear/clothing industry. A large part of the subcontracting cost started to be commissioned to subcontractors who were located outside the district, some of them in distant towns. Because of this growth, there was little expansion within the boundaries of the industrial district, because the new subcontractors were no longer located in the bordering towns. They worked

in different social-production contexts, where the conditions of the labour market and the productive features of the firms were different. Carpi became less and less a production area. The "head" of the industrial district (i.e. the independent firms) grew, while the productive "body" (i.e. the subcontractors) diminished and today it is not able to meet the production demand of the district. (Bigarelli and Crestanello, 1994, p. 137-38)

Bellandi (1994) emphasises private and public institution-making as a condition for the reproduction of dynamic industrial districts with growth potential. When difficulties with institution making or the support of local industrial policy arise in an industrial district, 'the basic conditions which sustain DIC are easily impaired, and the life-expectancy of such a district is relatively short' (Bellandi, 1994, p. 81). Such institution making is part of what Amin and Thrift (1994a) call 'institutional thickness', which they claim is of critical importance for 'the performance of local economies in a globalizing world' (Amin and Thrift, 1994a, p. v).

In this advocacy for a transition from the original 'industrial district Mark I' (i.e. districts without local government intervention) to 'industrial district Mark II' (i.e. districts with considerable government intervention) Brusco points out that 'industrial districts eventually face the problem of how to acquire the new technological capabilities which are necessary to revive the process of creative growth. It is here that the need for intervention appears' (Brusco 1990, p. 17). In another context, Brusco has claimed that 'industrial districts are slow to adopt new technologies, lack expertise in financial management, have little of the know-how required for basic research, and are unable to produce epoch-making innovations' (Brusco, 1992, p. 196). Along the same lines, Varaldo and Ferrucci (1996) maintain that 'the district firm presents a barrier to organizational innovation, if changes are linked to the assimilation of new competencies very different from the entrepreneur's technical culture. In this way, these firms have deficits in marketing and R & D activities' (p. 30).

Thus, most observers seem to agree that technological capabilities and the endogenous innovative capacity of industrial districts are important differentiating factors concerning their present development and of strategic importance for the future prospects of house districts (Asheim, 1994; Bellandi, 1994; Brusco, 1990; Crevoisier, 1994; Garofoloi, 1991b). Bellandi sees 'the assessment of the endogenous innovation capacities of the industrial districts ... as ... a key issue' (Bellandi, 1994, p. 73). More specifically this means the capability of SMEs in industrial districts to break path dependency and change technological trajectory through radical innovations. According to Varaldo and Ferrucci (1996), 'long-term strategic relationships, R&D investments, engineering skills, new technical languages and new organizational and inter-organizational models are need for supporting these innovative strategies in firms in industrial districts' (p. 32). Crevoisier (1994) emphasises the importance of understanding how industrial districts 'react to or generate radical innovations. Without making this point clear, it is not possible to make any prediction about the reproduction and the duration of such systems' (p. 259).

Figure 7.1 presents four ideal types of industrial districts with respect to technological capability-building (Asheim, 1994). Square I represents the original Marshallian model of an industrial district. However, the problem with these industrial districts is their relative low potential for endogenous technological capability-building; i.e. owing to the relatively low level of codified knowledge and technological know-how of SMEs within industrial districts, they are mainly able to adopt, adapt and develop incremental innovations. In Square II we find industrial districts with some potential for technological capability-building, due to the collective resources of the districts as they belong to the Mark II model, which to some extent compensates for the low level of internal resources and competence of the individual firm. Square II represents industrial districts with a good potential for technological capability-building due to a strong horizontal inter-firm cooperation normally found in these districts between firms with high levels of internal resources and competence. Last, Square IV is characterised by a high potential for technological capability-building due to the combined effect of the presence in the district of SMEs with high levels of internal resources and competence together with considerable public intervention. [6]

Thus, what is needed is an upgrading of industrial districts to 'the twenty-first century notion of an economy built on clusters of firms efficiently producing for high-quality, high-standard global markets on the basis of a regional innovation architecture' (Cooke, 1996, p. 61), as 'the systematic district strategy no longer sustains firms' competitive advantage. Rather this depends upon the internal capabilities of each firm or firm's network' (Varaldo and Ferrucci, 1996, p. 33). This new strategy can find support from modern innovation theory, originating from new institutional economics, which argues that 'regional production systems, industrial districts and technological districts are becoming increasingly important' (Lundvall, 1992, p. 3).

Modern innovation theory is developed as a result of criticism of the traditional dominant linear model of innovation as the main strategy for national R&D policies. The *linear model of innovation* was part of the Fordist era of industrial organisation and production, based on formal knowledge generated by R&D activity (codified scientific and engineering knowledge), large firms and national systems of innovation. Smith (1994) identifies the problem of this model along two dimensions. The first problem was 'an overemphasis on research (especially basic scientific research as the source of new technologies' (Smith, 1994, p. 2). Within this perspective a low innovative capacity could be explained by a low R&D activity. Consequently, technology policy in most western countries was directed towards increasing the level of basic research. The second problem was a 'technocratic view of innovation as a purely technical act: the production of a new technical device' (Smith, 1994, p. 2). The linear innovation model is thus, 'research-based, sequential and technocratic' (Smith, 1994, p. 2).

		Strong local co-operative environment	
		Industrial district Mark 1	**Industrial district Mark II**
SME's Internal resources and competence	**Low**	I Local production systems with <u>low</u> potentials for technological capability-building (Ex. Gnosjö, Sweden)	II Local production systems with <u>some</u> potentials for technological capability-building (Ex. Carpi and Reggio-Emilia in Emilia-Romagna)
	High	III Local production systems with <u>good</u> potentials for technological capability-building (Ex. Jæren, Norway; Sassuolo, Emilia-Romagna	IV Local production systems with <u>high</u> potentials for technological capability-building (Ex. Modena, Emilia-Romagna; Baden-Württtenberg, Germany

Figure 7.1 Typology of industrial districts with respect to innovative capability

Traditionally, the alternative of SMEs in order to upgrade their innovative capability has been to introduce (more) formal R&D-based product and process innovations. However, formal R&D activity has normally been out of reach for the majority of SMEs due to lack of financial as well as human resources.

However, 'it is now recognized that technological innovation and its contribution to economic growth is punctuated by discontinuities, nonappropriabilities, and processes of learning by doing, using and failing. Evolutionary theories of economic and technological change have now replaced the determinism of the linear model' (Felsenstein, 1994, p. 73). This criticism implies another and broader view of the process of innovation as a technical as well as a social process; as a non-linear process, 'involving not just research but many related activities' (Smith, 1994, p. 6), and as a process of interaction between firms and their environment (Smith, 1994). This implies a more sociological view on the process of innovation, in which interactive learning is looked upon as 'a fundamental aspect of the process of innovation' (Lundvall, 1993, p. 61). Lundvall emphasises that 'learning is predominantly an interactive and, therefore, a socially embedded process which cannot be understood without taking into consideration its institutional and cultural context' (Lundvall, 1992, p. 1). Also, Camagni emphasises that 'technological innovation ... is increasingly a product of social innovation, a process happening both at the intra-regional level in the form of collective learning processes, and through inter-regional linkages facilitating the firm's access to different, though localised, innovation capabilities' (Camagni, 1991, p. 8).

Consequently, Lundvall and Johnson (1994) use the concept of a 'learning economy' when referring 'first of all to the ICT (information, computer and telecommunication) - related techno-economic paradigm of the post-Fordist period. It is through the combination of widespread ICT-technologies, flexible specialisation and innovation as a crucial means of competition in the new techno-economic paradigm, that the learning economy gets firmly established' (p. 26). These perspectives on the 'learning economy' are based on the view that *knowledge* is the most fundamental resource in a modern capitalist economy, and *learning* the most important process (Lundvall, 1992), thus making the learning capacity of an economy of strategic importance to its innovativeness and competitiveness.

One of the consequences of the considerable more knowledge-intensive modern economies is that 'the production and use of knowledge is at the core of value-added activities, and innovation is at the core of firms' and nations' strategies for growth' (Archibugi and Michie, 1995, p. 1). Thus, in a learning economy 'technical and organisational change have become increasingly endogenous. Learning processes have been institutionalised and feed-back loops for knowledge accumulation have been built in so that the economy as a whole ... is "learning by doing" and "learning by using' (Lundvall and Johnson, 1994, p. 26).

Industrial districts in a post-Fordist 'learning economy'

Innovation and territorial agglomeration

The perspective of 'learning economies' and modern innovation theory emphasises that technological learning is a localised, and not a placeless, process (Storper 1995a). This is also supported by Porter, who argues that 'competitive advantage is created and sustained through a highly localized process' (Porter 1990, p. 19). According to Lazonick (1993), 'Porter argues that, ..., the building of a "home base" within a nation, or within a region of a nation, represents the organizational foundation for global competitive advantage' (p. 2). In contrast to this, Reich in his book *The Work of Nations* (1991) argues that 'the globalization of industrial competition has led to a global fragmentation of industry, thus making national industries and the national enterprises within them less and less important entities in attaining and sustaining global competitive advantage' (Lazonick, 1993, p. 2).

Thus, the major impact of Porter's (1990) book *The Competitive Advantage of Nations* is represented by a change in the understanding of the strategic factors which promote innovation and economic growth. Porter's main argument is that these factors are a product of localised learning processes, and that the importance of clusters is that they represent the material basis for an innovation based economy, which represents 'the key to the future prosperity of a nation' (Lazonick, 1993, p. 2). This argument is clearly based on Schumpeter's idea that 'competition in capitalist economies is not simply about prices, it is also a technological matter: firms compete not by producing the same products cheaper, but by producing new products with new performance characteristics and new technical capabilities' (Smith, 1994, p. 10). This is what Storper and Walker (1989) call 'strong competition' between quality-competitive' firms, which base their competitiveness on a 'differentiation' strategy, i.e. on innovative activity resulting in product innovations, in contrast to 'weak competition' between 'price-competitive' firms, i.e. firms which meet tougher competition with a search for cost (normally wage) and price reductions. Thus, while Porter focus on the importance of 'disembodied knowledge' in promoting innovativeness and competitiveness, Reich points at 'embodied knowledge', in the form of 'a work force that can find high-paid employment in the "global webs" of enterprise that are currently being spun around the world' (Lazonick, 1993, p. 2), as the most important factor in order to secure a nation's future prosperity [7]

Porter's cluster is basically an economic concept indicating that 'a nation's successful industries are usually linked through vertical (buyer/supplier) or horizontal (common customers, technology, channels, etc.) relationships' (Porter, 1990, p. 149). However, he emphasises that 'the process of clustering, and the interchange among industries in the cluster, also works best when the industries involved are geographically concentrated' (Porter, 1990, p. 157).

These ideas are more or less the same as those that Perroux, another Schumpeterian inspired economist, presented in the early 1950s. Perroux argued that it was possible to talk about 'growth poles' (or 'development poles') in an 'abstract economic space', i.e. firms which are linked with an innovative 'key industry' to form an industrial complex. According to Perroux, the growth potential and competitiveness of a growth pole could be intensified by territorial agglomeration (Haraldsen, 1994; Perroux, 1970).

Thus, the main argument for territorial agglomerations of economic activity in a contemporary capitalist economy is that they provide the best context for an innovation based economy. According to Amin and Thrift 'agglomerated learning capability becomes a condition for both dominating the relevant global economic networks *and* securing the cumulative industrial development of the "home base", by attracting and supporting the best quality domestic and overseas firms' (Amin and Thrift, 1995, p. 275).

The emphasis on interactive learning as a fundamental aspect of the process of innovation points to *cooperation* as an important strategy in order to promote innovations. The rapid economic development in the 'Third Italy', based on territorial agglomerated SMEs has drawn increased attention to the importance of cooperation between firms and between firms and local authorities in achieving international competitiveness. 'It is the success of the industrial districts in securing inter-firm cooperation and channelling the competitive forces towards such constructive ends of quality upgrading and technical change that brought them to the attention of the international research community' (You and Wilkinson, 1994, p. 276). According to Dei Ottati, 'this willingness to cooperate is indispensable to the realization of innovation in the ID which, due to the division of labour among firms, takes on the characteristics of a collective process. Thus, for the economic dynamism of the district and for the competitiveness of its firms, they must be innovative but, at the same time, these firms cannot be innovative in any other way than by cooperating among themselves' (Dei Ottati, 1994, p. 474).

Many observers have pointed to the importance of collaboration between territorial agglomerated firms in promoting international competitiveness. Pyke (1994) underlines close inter-firm cooperation and the existence of a supporting institutional infrastructure at the regional level (e.g. centres of real services) as the main factors explaining the success of Emilia-Romagna. Camagni points out that 'the collective learning processes that enhance the local creativity, the capability of product innovation and of "technological creation" ' (Camagni, 1991, p. 3), and You and Wilkinson (1994) are also of the opinion that 'a high degree of cooperation may be an important ingredient of industrial success' (p. 275).

Thus, if these observations are correct, this represents new 'forces' in the promotion of technological development in capitalist economies, implying a modification of the overall importance of competition between individual capitals. Of course, the fundamental forces in a capitalist mode of production constituting technological dynamism are still caused by the contradictions of the capital-capital relationship (Asheim, 1985). However, Lazonick argues, referring to Porter's empirical evidence (Porter 1990), that

> domestic cooperation rather than domestic competition is the key determinant of global competitive advantage. For a domestic industry to attain and sustain global competitive advantage requires continuous innovation, which in turn requires domestic cooperation. Domestic rivalry is an important determinant of enterprise strategies. But the substance of these competitive strategies - specifically whether they entail continuous innovation or cut-throat price-cutting - depends on how and to what extent the enterprises in an industry cooperate with one another. (Lazonick, 1993, p. 4)

Cooke (1994a) supports this view, emphasising that 'the co-operative approach is not infrequently the only solution to intractable problems posed by globalization, lean production or flexibilisation' (Cooke, 1994a, p. 32).

Through networking the ambition is to create 'strategic advantages over competitors outside the network' (Lipparini and Lorenzoni, 1994, p. 18). However, to achieve this it is important that networks are organised in accordance with the principle of 'strength of weak ties' (Granovetter 1973). Grabher argues that 'loose coupling with in networks affords favourable conditions for interactive learning and innovation. Networks open access to various sources of information and thus offer a considerably broader learning interface than is the case with hierarchical firms' (Grabher, 1993, p. 10).

Using this perspective on networks when discussing the relationship between competition and cooperation within industrial districts, competitive advantage is achieved internally through inter-firm cooperation and exploited externally through competition with firms of the 'outside' world. Lazonick argues that 'to fight foreign rivals requires a suspension of rivalry in order to build value-creating industrial and technological communities. Unless social organizations are put in place that can engage in innovation, heightened domestic rivalry will lead to decline' (Lazonick, 1993, p. 8).

The role of cooperation and competition and the innovative capacity of industrial districts

In the literature on industrial districts, from Piore and Sabel's book (1984) to the present, it has been emphasised that 'the central feature of the "industrial district" is the balance between competition and co-operation among firms' (You and Wilkinson, 1994, p. 259). Dei Ottati asserts that 'the cooperative elements contribute in a decisive way to the integration of the system, while forces of competition keep it flexible and innovative. This is because competition in the particular socio-economic district environment encourages better utilization of available resources and above all, development of latent capabilities and diffuse creativity' (Dei Ottati, 1994, p. 476). [8]

Porter (1990) also has similar problems in acknowledging the large and increasing influence of cooperation on the promotion of innovations and competitiveness. According to Porter, 'two elements - domestic rivalry and geographic industry concentration - have especially great power to transform the "diamond" into a system, domestic rivalry because it promotes upgrading of the entire national "diamond", and geographic concentration because it elevates and magnifies the interactions within the "diamond" ' (Porter, 1990, p. 131). He concludes 'that the most striking findings from our research ... is the prevalence of several domestic rivals in the industries in which the nation had international advantage. Rivalry has a direct role in stimulating improvement and innovation' (Porter, 1990, p. 143). Furthermore, Porter (1990) maintains that 'the broader effects of domestic rivalry are closely related to an old but often neglected notion in economics known as external economies' (p. 144).

This ambivalence regarding the relationship between, and the relative importance of, competition and cooperation is basically caused by a traditional, Marshallian perception of industrial districts and of the achievement of external economies through vertical cooperation (Asheim, 1996).[9] In my view, one of the constraining factors in moving beyond the domination of incremental innovations in industrial districts is the fierce competition between subcontractors specialising in the same products or phases of production, and vertically linked to the commissioning firms. This limits the potential for horizontal technological cooperation (Asheim, 1994). In addition, a characteristic of industrial districts is that they are made up of independent small firms with no single big firm acting as a centre for strategic decision-making. The problem, in this respect, is that owing to a shortage of both human and financial resources SMEs lack the capability to build up and support a necessary level of research and development capacity in accordance with the linear model of innovation, and, thus, should more systematically exploit the broader social basis of innovative activity.[10]

This means that the possibilities and potentials of the 'learning economy' are not fully recognised. In this connection an important aspect of a 'learning economy' is that 'the organisational modes of firms are increasingly chosen in order to enhance learning capabilities: networking with other firms, horizontal communication patterns and frequent movements of people between parts and departments, are becoming more and more important' (Lundvall and Johnson, 1994, p. 26).

This can be illustrated by an example from the Third Italy, where a firm started to cooperate with its suppliers in developing new products (i.e. product innovations) in order to institutionalise a continual organisational learning process. This cooperation played a central role in shortening the product cycle, improving the product quality and increasing the competitiveness of the firm (Bonaccorsi and Lipparini, 1994). The firm redefined its relations to its major suppliers based on the recognition that 'a network based on long-term, trust-based alliances could not only provide flexibility, but also a framework for joint learning and technological and managerial innovation. To be an integral partner in the development of the total product, the supplier must operate in a state of constant learning, and this process is greatly accelerated if carried out in an organizational environment that promotes it' (Bonaccorsi and Lipparini, 1994, p. 144).

Generally, Semlinger notes that 'modern purchasing policies are heading towards an intensification of interim cooperation referred to as "integration of suppliers" or "dissolving the boundaries of the enterprise" ' (Semlinger, 1993, p. 169). Also Tödtling (1995) notes that 'subcontracting relationships ... have changed substantially: they are no longer confined to the goal of cost savings only, but increasingly include aspects of product quality and technology development and improvement. This implies more selective and fewer but stronger relationships between firms since they cover not just production but also quality control, joint research and development as well as information exchange on and coordination of future planning' (p. 14). This represents an understanding of flexibility as primarily a function of the innovative capability of firms and districts, i.e. a more dynamic perspective than the traditional focus on internal and external flexibility caused by new computerised production equipment and vertical disintegration.

Forms of inter-firm cooperation

The importance of horizontal inter-firm cooperation with respect to promoting innovations highlights the qualitative aspects of networking, i.e. specifically the governance structure of the networks. Normally, the basis of inter-firm cooperation between manufacturing firms is a production system, understood as 'a complex input-output system of linked production chains with vertical, horizontal, and diagonal links'(Dicken, 1994, p. 103), i.e. technologically and organisationally integrated production units in the manufacture of final products (Sheard, 1983).

However, 'a *production system* is much more than the input-output systems ... Any real production system involves an input-output system set in a context of relations of power and structures of decision-making, which we call governance. The nature and dynamics of given input-output systems cannot be understood without taking into account the phenomenon of governance' (Storper and Harrison, 1990, p. 10). Thus, a governance structure refers in general to 'he degree of hierarchy and leadership (or their opposites, collaboration and cooperation)' in a production system (Storper and Harrison 1990, p. 10).

The new ways of organising industrial production can take various forms. The specific new form of industrial organisation resulting from close inter-firm networking is represented by 'quasi-integration' (Leborgne and Lipietz, 1988). Quasi-integration refers to relatively stable relationships between firms, where the principal firms (i.e. the buyers) aim at combining the benefits of vertical integration as well as vertical disintegration in their collaboration with suppliers and subcontractors (Haraldsen, 1995). According to Leborgne and Lipietz (1992) 'quasi-integration minimizes both the costs of coordination (because of the autonomy of the specialized firms or plant), and the cost of information/transaction (because of the routinized just-in-time transactions between firms). Moreover the financial risks of R & D and investments are shared within the quasi-integrated network' (p. 341).

Leborgne and Lipietz (1992) distinguish between three different forms of quasi-integration. The most extreme case is called 'vertical quasi-integration', where 'the buyer has at its disposal the know-how of the subcontractor' (Leborgne and Lipietz, 1992, p. 341). By contrast, there is the case of 'horizontal quasi-integration', when 'partnership and strategic alliance link a supplier with specific technology to a regular customer of another sector of the division of labor' (Leborgne and Lipietz, 1992, p. 341). The general case is, however, the intermediate situation of 'oblique quasi-integration', where the customer orders 'specific goods which are part of the process of production' (Leborgne and Lipietz, 1992, p. 342), but where the supplier "is fully responsible for the process of production" (Leborgne and Lipietz, 1992, p. 342).

Haraldsen (1995) maintains that the relations between 'differentiated suppliers' and buyers are characterised by 'horizontal quasi-integration'; the relation between 'specialised suppliers' and buyers by 'oblique quasi-integration'; while 'vertical quasi-integration' corresponds to the relations between capacity subcontractors and their principal firms.

'Vertical quasi-integration' represents an externalisation of the technical division of labour from a unit of production (i.e. a factory) to a production system. Such networks are characterised by an intensive 'weak' competition between potential subcontractors and, consequently they are not very innovative.

In contrast, 'oblique quasi-integration' can typically be constituted by 'user-producer' relations which imply potentially good opportunities for 'learning by interacting', based on an externalisation of the technical division of labour within a production system. As interactive learning is at the core of the process of innovation in a 'learning economy', this form of inter-firm cooperation represents important possibilities for carrying out radical product and process innovations.

'Horizontal quasi-integration' represents a form of inter-firm cooperation, where the supplier delivers goods which are not an integrated part of the production process of the principal firm, as opposed to the two previous categories. In this case, it is more a question of complementary products or services, which imply that such horizontal relations are based on a social division of labour (Haraldsen, 1995).

Leborgne and Lipietz (1992) maintain that the more horizontal the ties between the partners in the network are (i.e. networks dominated by oblique or horizontal quasi-integration), the more efficient the network as a whole is. This is also emphasised by Håkansson, who points out that 'collaboration with customers leads in the first instance to the step-by-step kind of changes (i.e. incremental innovations), while collaboration with partners in the horizontal dimension is more likely to lead to leap-wise changes (i.e. radical innovations)' (Håkansson, 1992, p. 41). Generally Leborgne and Lipietz (1992) argue that 'the upgrading of the partner increases the efficiency of the whole network' (p. 399). Cooke (1996) refers to research on innovative SMEs in the UK showing that, in general, 'innovative SMEs differ from un-innovative SMEs by having dense external networks involving other innovative SMEs in a variety of technical, marketing and manufacturing relationships involving infra structural institutions such as universities and private sector research institutes' (p. 55).

This reorganisation of networking between firms can be described as a change from a domination of vertical relations between principal firms and their subcontractors based on a technical division of labour to horizontal relations between principal firms and suppliers based on a social division of labour. Patchell refers to this as a transformation from production systems to learning systems, which implies a transition from 'a conventional understanding of production systems as fixed flows of goods and services to dynamic systems based on learning' (Patchell, 1993, p. 797). Patchell argues that Japan has developed 'a social technology that resolves the transaction cost trade-offs ... between internal and external governance structures' (Patchell, 1993, p. 797). This social technology can be conceptualised by using

Asanuma's reformulation of Williamson's static concept of 'asset-specific relation' to the more dynamic concept of 'relation-specific skill' (Asanuma, 1989). According to Patchell (1993), relation-specific skill is 'the crux for comprehending the shift from production systems to learning systems' (Patchell, 1993, p. 797). Such a transformation requires 'organisational integration', which is a result of the organisational capability of firms (Lazonick and Smith, 1995). Organisational integration is manifest when 'the relationships among participants in a specialized division of labour permit their activities to be planned and co-ordinated to achieve specified goals' (Lazonick and Smith, 1995, p. 9).

'Relation-specific skill' is defined as 'the skill required on the part of the supplier to respond efficiently to the specific needs of a core firm. Formation of this skill requires that learning through repeated interactions with a particular core firm be added to the basic technological capability which the supplier has accumulated' (Asanuma, 1989, p. 28). On this basis the distinction is made between 'design-supplied' (DS) and 'design-approved' (DA) suppliers, which is constituted by the difference between 'the technological dependency of DS firms on the core firm to design the product that they produce' (Patchell, 1993, p. 811), and DA firms, which 'will exchange technological information on an equal level of sophistication with core firms, and these DA suppliers serve regional, national, and international markets' (Patchell, 1993, p. 812). The qualitative difference between these two forms of inter-firm cooperation is represented by the difference between vertical relations of core firms and subcontractors (i.e. a DS firm) on the one hand, and horizontal relations of core firms and suppliers (i.e. DA firms) on the other hand, where 'core firm and supplier share information and participate in an evolving learning and creative process' (Patchell, 1993, p. 814).[11] Thus, the promotion of horizontal inter-firm cooperation must have a central role in the future industrial policy.

'Learning economies' and organisational and institutional innovations

A contrast to traditional Marshallian industrial districts is the decreased importance of 'the collectivist and institutional basis for successful co-ordination' (You and Wilkinson 1994, p. 265). According to You and Wilkinson, Marshall meant that 'the role of employers' and workers' organizations and the state was limited. By contrast, in recent discussions of industrial districts, collectivity in the form of direct inter-firm relationships, formal and informal institutions and public policy play a central role in establishing and guaranteeing business and labour standards, fostering innovations and technology diffusion and organizing education and training' (p. 266). This is in accordance with a 'learning economy' in which 'a wide array of institutional mechanisms can play a role' (Morgan, 1995, p. 6). Thus, generally speaking 'institutional characteristics of the learning economy becomes a crucial question' (Lundvall and Johnson, 1994, p. 30).

Furthermore, Lundvall and Johnson emphasise that 'the firms of the learning economy are to a large extent "learning organisations" '(Lundvall and Johnson, 1994,

p. 26). A dynamic flexible 'learning organisation' can be defined as one that promotes the learning of all its members and has the capacity of continuously transforming itself by rapidly adapting to changing environments by adopting and developing innovations (Pedler et al, 1991; Weinstein, 1992). Thus, important organisational and institutional innovations in a 'learning economy' are the formation of 'learning organisations' not only at an intra-firm, but also at an inter-firm, level as well as at a district of regional level.

Intra-firm cooperation

Lundvall and Johnson (1994) argue that 'the firm's capability to learn reflects the way it is organised. The movement away from tall hierarchies with vertical flows of information towards more flat organisations with horizontal flows of information is one aspect of the learning economy' (p. 39). This is in line with Scandinavian experiences, which have shown that flat and egalitarian organisations have the best prerequisites of being flexible and learning organisations, and that industrial relations characterised by strong involvement of functional flexible, central workers is important in order to have a working 'learning organisation'. Such organisations will also result in well functioning industrial relations, where all the employees (i.e. the (skilled) workers as well as the managers) will have a certain degree of loyalty towards the firm. Experience shows that 'the process of continuous improvement through interactive learning and problem-solving, a process that was pioneered by Japanese firms, presupposes a workforce that feels actively committed to the firm' (Morgan, 1995, p. 11).

Brusco - with special reference to the industrial districts of Emilia-Romagna - points to the dominating model of production in the districts 'that was able to be efficient and thus competitive on world markets, in which efficiency and the ability to innovate were achieved through high levels of worker participation and were accompanied by working conditions that were acceptable' (Brusco, 1995, p. 5). In general, Porter points out that 'differences in managerial approaches and organizational skills create advantages and disadvantages in competing in different types of industries. Labor-management relationships are particularly significant in many industries because they are so central to the ability of firms to improve and innovate' (Porter, 1990, p. 109).

According to modern organisational theory and practice the challenges of the 'learning economy' are increasingly being institutionalised as firm-internal 'development organisations'. They represent the framework for carrying out the process of continuous improvement in productivity and competitiveness. The strategy behind such an organisational innovation is to make 'labour productivity "endogenous" and raise it above market levels, hence not transferable to other firms' (Perulli, 1993, p. 110).

A strong and broad involvement within an organisation will also make it easier to use and diffuse informal or 'tacit', non-R&D based knowledge, which in a 'learning economy' has a more central role to play in securing continuous innovation.

161

'Transactions' with 'tacit' knowledge within and between networking organisations require trust, which is easier to establish and reproduce in flat organisations than in hierarchical ones. In a study of successful intra-firm reorganisation of SMEs in Baden-Württenmberg Herrigel reports that one company 'set out to constitute "trusting" relations among all actors within the firm, regardless of role or position in the organization, which were informed by mutual respect. It discouraged thinking in terms of hierarchy and status and made all information about the company available to everyone within it' (Herrigel, 1996, p. 46). According to Lipparini and Lorenzoni (1994) 'a high dose of trust serves as substitute for more formalised control systems' (p. 18); see also Lorenz, 1992, and Sabel, 1992). In organisations characterised by an authoritarian management style the attitude of the employees will often be to keep 'the relevant information to themselves' (You and Wilkinson, 1994, p. 270).

Inter-firm cooperation Generally, inter-firm networking is of strategic importance to SMEs due to their lack of financial and human resources and/or marketing capabilities, which restrict their innovative capacity. According to Brusco, it is 'the fact of being a "system" rather than being a "single firm" that defines the degree of sophistication of these industrial structures' (Brusco, 1986, p. 194). In this way, the internal skill and competence of firms are strengthened through inter-firm collaboration and can, furthermore, be supported by local structures outside the firm. This strategy could be characterised as 'learning-by-interacting', of which the interactions between producers and users of intermediate products and between suppliers and users of machine tools and business services represent the main forms of cooperation. Herrigel refers to cases of successful adjustment and restructuring of SMEs in Baden Württemberg, where 'relations with suppliers ... were intensified so that important providers were drawn directly into the development process' (Herrigel, 1996, p. 47). Such cooperation can result in a largely improved innovative capacity of SMEs within industrial districts. Russo concludes her analysis of technological development in Sassoulo, Emilia-Romagna, by underlining the importance of 'the interrelationships between firms and their proximity to each other. Together these provide the basis for the process of generation and adoption of new techniques' (Russo, 1989, p. 215).

The spatial proximity of interacting firms is an important enabling factor in stimulating inter-firm 'learning networks' involving long-term commitment. Håkonsson (1992) claims that 'the importance of proximity is particularly noticeable in horizontal relationships, but it is not altogether absent in the case of vertical relations' (p.125). In this way, the ability to generate 'new knowledge by combining internal and external learning could then be a critical variable in understanding SME's innovative capabilities' (Lipparini and Sobrero 1994, p. 136). [12]

However, this may require a change in industrial organisation towards a more hierarchical group-formation of firms, which can be observed in several industrial districts in the 'Third Italy' (Zeitlin, 1992). Sometimes these new groups are controlled by larger companies outside the industrial districts, but most commonly the groups are formed by SMEs under competitive pressure in order to stay competitive (Zeitlin, 1992). Cooke and Morgan (1994) maintain that 'new corporate hierarchies appear to be emerging in the region's industrial districts as a result of a growing concentration of capital through especially mergers and takeovers' (p. 106). In addition to providing SMEs with financial and human resources to increase the innovative capacity in order to improve their international competitiveness, the formation of groups can be a strategy for establishing more systematic horizontal inter-firm networking promoting technological cooperation. According to Cooke 'recent evidence from ... the Third Italy, suggests group-formation has enabled firms in industrial districts to outperform their sector generally' (Cooke, 1994b, p. 24). Furthermore, in this perspective the organisation of innovation network between industrial districts and the external world (Camagni, 1991), giving priority to horizontal inter-firm technological cooperation to ensure the adoption and diffusion of radical innovations, is very important.[13]

In this process of restructuring in order to survived, 'the small firms network will have to modify some of its basic characteristics and evolve toward a more concentrated structure, in which a limited number of leading firms or firm associations will perform a key role in both strategic and commercial terms' (Malerba, 1993, p. 257). This is a characteristic tendency of the development of industrial districts in the beginning of the 1990s, and will have a decisive effect on the future prospects of this form of territorial based industrial agglomerations.

Regional cooperation

An important innovation in the institutional set-up of 'learning regions' would be the establishment of territorial embedded regional systems of innovation (Asheim, 1995) - which could improve what has been called 'systemic innovation' with reference to Baden-Württemberg (Cooke and Morgan, 1994) - as a strategic part of a regional innovation policy. The aim of such a policy is through public intervention to support organisational innovations such as 'centres of real services' in the industrial districts of Emilia-Romagna (Brusco, 1992), which have turned out to be successful in modernising the economic structure of the districts and, thus, have strengthened their competitive advantage. In general, Amin and Thrift (1995) emphasise 'the need for enterprise support systems, such as technology centres or service centres, which can help keep networks of firms innovative' (p. 12). And, according to Cooke, 'the *region* is a most appropriate economic and administrative entity around which to plan networking approaches' (Cooke, 1994a, p. 33).

The need for such public intervention could be illustrated with reference to what the GREMI-group calls 'innovative milieu', i.e. 'The set, or the complex network of mainly informal social relationships on a limited geographical area, often determining a specific external "image" and a specific internal "representation" and sense of belonging, which enhance the local innovative capability through synergetic and collective learning processes' (Camagni, 1991, p. 3). In this perspective, creativity and continuous innovation is considered to be a result of 'a collective learning process, fed by such social phenomena as intergenerational transfer of know-how, imitation of successful managerial practices and technological innovations, interpersonal face-to-face contacts, formal or informal cooperation between firms, tacit circulation of commercial, financial or technological information' (Camagni, 1991, p. 1).

However, the basic problem with the 'innovative milieu' approach is that, beyond referring to 'industrial atmosphere' and different forms of incremental innovations, it does not specify the mechanisms and processes which promote innovative activity more successfully in some regions than in others, i.e. 'Why localization and territorial specificity should make technological and organizational dynamics better' (Storper, 1995b, p. 203). Their focus is too much on what they call the 'territorial logic' of development processes, which on the one hand misses the central point of the 'productive' balance of the functional and territorial modes of integration, which has been the key to the industrial and economic success of the industrial districts (Asheim, 1992; 1994), without fully understanding the challenges of the 'learning economy' on the other hand.[14]

Furthermore, the strong focus on the advantages of the territorial mode of integration increases the possibilities of ignoring the danger of supporting economic and social structures which create 'lock-in' situations through the 'weakness of strong ties' (Granovetter, 1973), which often characterises old industrial agglomerations of SMEs (Glasmeier, 1994). Porter (1990) argues that 'geographic concentration does carry with it some long-term risks, however, especially if most buyers, suppliers and rivals do not operate internationally' (p. 157). And Grabher (1993) points out what he calls an 'embeddedness dilemma' with respect to major social, economic and technological changes. However, Camagni (1991) warns about such development tendencies when he maintains that 'innovation networks and cooperation agreements become the strategic instruments that local environments may utilize in order to avoid an "entropic death" which always threatens too closed systems, and to keep on exploiting at the same time the advantages provided by their internal synergies, their industrial "memory" and atmosphere' (p. 5).

The challenge of 'learning regions' is to increase the innovative capability of SME-based industrial agglomerations through identifying 'the economic logic by which milieu fosters innovation' (Storper, 1995b, p. 203). At the regional level this points to the importance of disembodied technical progress, i.e. progress 'which can occur independently of changes in physical capital stock' (de Castro and Jensen-Butler, 1993, p. 1), and 'untraded interdependencies', i.e. 'A structured set of technological externalities which can be a *collective asset* of groups of firms/industries within countries/regions' (Dosi, 1988, p. 226), with respect to establishing regional systems of innovation, together with territorial embedded Marshallian agglomeration economies (i.e. 'An atmosphere cannot be moved' (Marshall, 1919, p. 284)), disembodied technical knowledge and 'untraded interdependencies' constitute the material basis for the formation of territorial embedded regional systems of innovation as an alternative to regionalised national systems of innovation represented by science parks and other top-down technology policies based on the linear model of innovation (Asheim, 1995; Henry et al, 1995).

Marshall (1919) underlines that 'a man can generally pass easily from one machine to another; but the manual handling of a material often requires a fine skill that is not easily acquired ...: for that is characteristic of a special industrial atmosphere' (Marshall, 1919, p. 287). According to de Castro and Jensen-Butler (1993) 'rapid disembodied technical progress requires ... a high level of individual technical capacity, collective technical culture and a well-developed institutional framework ... [which] ... are highly immobile in geographical terms' (p. 8). Finally, Dosi argues that 'untraded interdependencies' represent 'context conditions' which generally are country- or region-specific, and of fundamental importance to the innovative process (Dosi 1988, p. 226; see also Storper, 1995b, 1995c). Amin and Thrift (1994b) precisely underline 'the role of localised "untraded interdependencies" in securing learning and innovation advantage in inter-regional competition' (p. 12).

Conclusion: Industrial districts as 'learning regions'

In this chapter I have discussed the prospects of endogenous regional development in the context of strong globalisation tendencies. I agree that 'there exists a viable dynamic, competitive and socially desirable paradigm of small and medium-sized enterprise development, following the principle of the Italian industrial districts prototype' (Lyberaki and Pesmazoglou, 1994, p. 509), conditioned by a transformation of districts into 'learning regions'.

However, the ultimate question still remains: how much long-term strategic planning can be introduced into the decentralised industrial systems of SMEs and be undertaken by new corporate leaders without totally destroying the innovative milieu, flexibility and consequent economic dynamism of industrial districts? Dimou (1994) also asks if it is possible 'to reconcile and industrial dynamic with rapid rhythms of change and a social dynamic which evolves in a rather slow and progressive way without exposing the whole system to centrifugal pressures leading to disintegration' (Dimou, 1994, p. 27).

The viability, dynamism and competitiveness of SMEs in industrial districts has basically been the result of the way 'flexible specialization works by violating one of the assumptions of classical political economy: that the economy is separate from society. Markets and hierarchies ... both presuppose the firm to be an independent entity. ... By contrast, in flexible specialization it is hard to tell where society ends, and where economic organization begins' (Piore and Sabel, 1984, p. 275).

However, as emphasised by Bellandi (1994) and Brusco (1990) it is a question of a potential collective and innovative capacity of territorial agglomerated SMEs, which has to be systematically developed and supported both at the intra-firm, the inter-firm and the district or regional level. This perspective emphasises the importance of organisational innovations to promote cooperation, primarily through

165

the formation of dynamic flexible learning organisations within firms in network and between firms and society regionally. Such learning organisations must be based on strong involvement at the intra-firm level, on horizontal cooperation at the inter-firm level, and on the embeddedness of regional systems of innovation at the regional level. This could, together with other necessary organisational and social innovations in the regional institutional set-up, contribute to turning industrial districts into 'learning regions'.

Such 'learning regions' would be in a much better position than 'traditional' industrial districts to avoid a 'lock-in' of development caused by localised path-dependency. In a 'learning economy' the competitive advantage of firms and regions is based on innovations, and innovation processes are seen a socially and territorially embedded, interactive learning processes. Based on modern innovation theory it could be argued that SMEs in industrial districts can develop a large innovative capacity and thus a new competitive advantage.

This new understanding of the institutional and cultural context of a 'learning economy' stresses the importance of the 'fusion' of the economy with society, where socio-cultural structures and other broader historical factors are not only looked upon as reminiscences from pre-capitalist civil societies, but a necessary prerequisites for regions in order to be innovative and competitive in a post-Fordist global economy. This forces a re-evaluation of 'the significance of territoriality in economic globalisation' (Amin and Thrift 1995, p. 8).

In this way 'learning regions' could have the possibilities of transcending the contradictions between functional and territorial integration through a new regionalised integration of the traditional 'contextual' knowledge of industrial districts and the 'codified' knowledge of the global economy within the framework of territorial embedded regional systems of innovation.[15]

166

Notes

1.   When looking at industrial districts 'what is relevant is no longer the characteristics of one single firm, but the characteristics of the industrial district of which the small firm is a part' (Brusco, 1986, p. 187).

2.   Only when the process of vertical disintegration takes the form of 'horizontal quasi-integration' it is a question of a broadening of the social division of labour (Haraldsen, 1995; Leborgne and Lipietz, 1992).

3.   Even if everyone seems to know the basic distinction by Marx between the social and the technical division of labour, i.e. the first referring to the production of commodities (use-values) for the market and the second referring to functions (tasks) in a production process carried out by individual workers inside a factory or workshop (i.e. internal functional specialisation), problems start as soon as the concepts are used in analyses of the contemporary capitalist world economy. First, there seems to be a language problem as English only has one word 'social' to give meaning to the two German words 'gesellschaftlich' and 'sozial'. When Marx talked about the 'social division of labour' he was referring to 'gesellschaftliche' and not to 'soziale' Arbeitsteilung (division of labour). The confusion caused by the differences in languages can be illustrated by this statement from Sayer (1992) : 'There is nothing asocial about the technical division of labour; it is responsive to social differences such as gender divisions, as well as to technical influences' (p.347; see also Sayer and Walker, 1992). In this connection it is important to remember that the technical division of labour is the specific capitalist form of division of labour, developed as a product of the capitalist social relations of production first manifest in manufacture (thus, also called the manufacturing division of labour by Marx), and consequently responsible for the social stratification in capitalist societies. In contrast, the social division of labour has existed since the establishment of markets. In fact, there is basically more 'social' about the hierarchy of workers in the production process (i.e. the technical division of labour) than about the differences between an artisan making shoes and another making cloth (i.e. the social division of labour). Secondly, the technical division of labour still seems normally to be perceived only as an internal division of labour (i.e. inside a production unit), as is the case when Storper and Harrison (1990) refer to 'labor relations inside the production unit, the detail division of labor' (p. 28). When doing this they miss an important developmental tendency in the capitalist world economy, the externalisation of the technical division of labour outside the traditional boundaries of the firm, representing one of the most dynamic elements in the formation of local and global production systems. However, one of the few observers who has noticed this is Sheard (1983) in his study of the

auto-production systems in Japan. Here he points out that 'the just-in-time system is an extension of the principles of the Ford conveyor belt system of factory production to the regional production system of assembly plants and subcontractors. ... The timing and specialisation which characterise auto-production systems in Japan are analogous to those achieved in modern factory production' (Sheard, 1983, p. 62 and 64).

4.   Furthermore, disembeddedness should not be confused with the undersocialised view of neoclassical economics as Sally (1994) indicates, when he claims that 'disembeddedness obtains in economies that approximate the neoclassical conception of arm's length transactions between different economic actors in functioning, freely competitive markets' (Sally, 1994, p. 169). However, disembeddedness is simply a question of an economy which is *functionally* integrated in contrast to a *territorially* integrated, embedded economy.

5.   However, I do not disagree with the general conclusion of Porter (1990), when he maintains (as referred to by Sally (1994)) that 'the globalization of industries does not destroy national differences, on the contrary, the unique competitive advantages of the home base of the MNE continue to be at the vortex of, and launching pad for, global strategies and provide the added edge for firms in international competion' (Sally, 1994, p.181). In the same way it may be true that 'for some MNEs, the regional level *within* nation states is very important, with tight links to local governments and surrounding clusters of local suppliers, subcontractors, educational institutes and research centres' (Sally, 1994), p.184). A good example of this would be Baden-Württemberg with TNCs as Mercedes-Benz and Bosch (Cooke and Morgan, 1994)

6.   For a more detailed analysis of the technological capability-building of these different (ideal) types of industrial districts, see Asheim (1994).

7.   For a more elaborated discussion of the concepts of 'embodied' and 'disembodied' knowledge, see the section below on Regional Cooperation.

8.   This assertion obviously weakens her earlier statement (page 474).

9.   Porter has an explicit reference to Marshall (in a footnote) when he discusses the relation between domestic rivalry and external economies.

10.   In addition, the emphasis on the importance of domestic rivalry and competition in influencing factor creation (Porter, 1990), could reflect the survival of the view of 'orthodox economics in which cooperation is

regarded exclusively as an attempt to distort prices and is therefore inefficient' (You and Wilkinson 1994, p. 275).

11. In a comparison of the typology of Patchell with the one of Leborgne and Lipietz, 'design-supplied' subcontractors correspond to 'vertical quasi-integration' while 'design-approved' suppliers can be represented by 'oblique quasi-integration' as well as 'horizontal quasi-integration'.

12. Camagni points out that such 'interfirm networks may enrich the respective territorial environments or "milieux" through the opportunities they provide for information interchange, explicit or tacit know-how transmission, and skilled factors mobility through the networks' (Camagni, 1991, p. 5).

13. According to Camagni the formation of innovation networks with external and specialised milieus may provide local firms with 'the complementary assets they need to proceed in the economic and technological race' (Camagni, 1991, p. 4).

14. Dimou argues along the same lines that 'the industrial districts appears as an organizational fact, stemming from the interactions between an industrial dynamic defined at a global level and a social dynamic defined at a territorial level. As long as these two components of the district evolve in the same way - that is, as long as the territory regulates efficiently the industrial process - the district structure subsists through time'. (Dimou, 1994, p. 28).

15. These new views of the workings of a modern economy are also shared by social economics (or 'socio-economics') (Amin and Thrift, 1994b) and the 'embeddedness' approach (Grabher, 1993).

References

Amin, A. (1993), 'The globalizations of the economy. An erosion of regional networks?', in Grabher, G. (ed.), *The Embedded Firm*, Routledge, London and New York, pp. 278-295.

Amin, A. and Thrift N. (eds) (1994a), *Globalization, Institutions, and Regional Development in Europe* Oxford University Press, Oxford.

Amin, A. and Thrift N, (1994b), 'Institutional issues for the European regions : from markets and plans to socioeconomics and power of association. *Economy and Society.*

Amin, A. and Thrift, N. (1995), 'Territoriality in global political economy', *Nordisk Samhällsgeografisk Tidskrift,* vol. 20, pp. 3-16.

Archibugi, D. and Michie, J. (1995), 'Technology and innovation: an introduction', *Cambridge Journal of Economics,* vol. 19, pp. 1-4.

Asanuma, B. (1989), 'Manufacturer-supplier relationships in Japan and the concept of the relation-specific skill', *Journal of the Japanese and International Economies,* vol. 3, pp. 1-30.

Asheim, B.T. (1985), 'Capital accumulation, technological development and the spatial division of labour: A framework for analysis', *Norwegian Journal of Geography,* vol. 45, no. 4, pp. 87-97.

Asheim, B.T. (1992), 'Flexible specialisation, industrial districts and small firms: a critical appraisal', in Ernste, H. and Meier, V. (eds), *Regional development and contemporary industrial response. Extending flexible specialisation*, Belhaven Press, London, pp. 45-63.

Asheim, B.T. (1994), 'Industrial districts, inter-firm co-operation and endogenous technological development: the experience of developed countries', in *Technological Dynamism in Industrial Districts: An Alternative Approach to Industrialization in Developing Countries?, UNCTAD,* United Nations, New York and Geneva, pp. 91-142.

Asheim, B.T. (1995), 'Regionale innovasjonssystem - en sosialt og territorielt forankret teknologipolitikk?', *Nordisk Samhällsgeografisk Tidskrift,* vol. 20, pp. 17-34.

Asheim, B.T. (1996), 'Industrial districts as "learning regions": a condition for prosperity?,' *European Planning Studies,* vol 4, no. 4 (forthcoming).

Becattini, G. (1990), 'The Marshallian industrial district as a socio-economic notion', in Pyke, F., Becattini, G. and Sengenberger, W. (eds), *Industrial Districts and Inter-firm Co-operation in Italy,* International Institute for Labour Studies, Geneva, pp. 37-51.

Bellandi, M. (1989), 'The industrial district in Marshall', in Goodman E. and Bamford J. (eds), *Small Firms and Industrial Districts in Intaly,* Routledge,London, pp. 136-52.

170

Bellandi, M. (1994), 'Decentralized industrial creativity in dynamic industrial districts', in *Technological Dynamism in Industrial Districts: An alternative approach to Industrialization in Developing Countries'* ?, UNCTAD, United Nations, New York and Geneva, pp. 73-87.

Bianchi, P. and Giordani, M.G. (1993), 'Innovation policy at the local and national levels: The case of Emilia-Romagna', *European Planning Studies,* vol. 1, no. 1, pp. 541.

Bigarelli, D. and Crestanello, P. (1994), 'An analysis of the changes in the knitwear/clothing district of Carpi during the 1980s', *Entrepreneurship & Regional Development,* vol. 6, pp. 127-144 .

Bonaccorsi, A. and Lipparini, A. (1994), 'Strategic partnerships in new product development' An Italian case study, *Journal of Product Innovation Management,* vol.11, no. 2, pp. 135-46.

Brusco, S. (1986), 'Small firms and industrial districts: The experience of Italy' in Keeble, D. and Wever, E. (eds), *New Firms and Regional Development in Europe,* Croom Helm, London, pp. 184-202.

Brusco, S. (1990), 'The idea of the industrial district: its genesis', in Pyke, F., Becattini, G. and Sengenberger, W. (eds), *Industrial Districts and Inter-firm Co-operation in Italy*, International Institute for Labour Studies, Geneva, pp. 10-19.

Brusco, S. (1992), 'Small firms and the provision of real services' in Pyke, F. and Sengenberger, W. (eds), *Industrial Districts and Local Economic Regeneration,* International Institute for Labour Studies, Geneva, pp. 177-96.

Camagni, R. (1991), 'Introduction: from the local "milieu" to innovation through cooperation networks', in Camagni, R. (ed.), *Innovation Networks: Spatial Perspectives.* Belhaven Press, London, pp. 1-9.

Castro, E. de and. Jensen-Butler, C. (1993), *Flexibility, Routine Behaviour and the Neo-classical Model in the Analysis of Regional Growth,* Institute of Political Science, Aarhus.

Cooke, P. (1994a), 'The co-operative advantage of regions', paper prepared for Centenair Harald Innis Centenary Celebration Conference on "Regions, Institutions, and Technology: Reorganizing Economic Geography in Canada and the Anglo-American World", University of Toronto, September.

Cooke, P. (1994b), 'The Baden-Württemberg machine tool industry: Regional responses to global threats', paper presented at the Workshop of the Centre of Technology Assessment in Baden-Württemberg on "Explaining regional competitiveness and the capability to innovate - the case of Baden-Württemberg" Stuttgart, June.

Cooke, P. (1996), 'Building a Twenty-first Century Regional Economy in Emilia-Romagna', *European Planning Studies,* vol. 4, no. 1, pp. 53-62.

171

Cooke, P. and Morgan, K. (1994), 'Growth regions under duress: renewal strategies in Baden-Württemberg and Emilia-Romagna', in Amin, A. and Thrift, N. (eds), *Globalization Institutions, and Regional Development in Europe,* Oxford University Press, Oxford, pp. 91-117.

Crevoisier, O. (1994), Book review (of Benko, G. and Lipietz A. (eds), Les regions qui gagnent, Paris, 1992). *European Planning Studies, vol.* 2, no.2, pp. 258-60.

Dei Ottati, G. (1994), 'Cooperation and competition in the industrial district as an organization model', *European Planning Studies, vol.* 2, no. 4, pp. 463-83.

Dicken, P. (1994), 'Global-Local Tensions: Firms and States in the Global Space-Economy', *Economic Geography,* vol. 7, no. 2, pp. 101-128.

Dicken, P. et al (1994), 'Strategies of transnational corporations and European regional restructuring: some conceptual bases', in Dicken, P. and Quevit, M. (eds), *Transnational Corporations and European Regional Restructuring,* Netherlands and Geographical Studies 181, The Royal Dutch Geographical Society/Faculty of Geographical Sciences, Utrecht University, Utrecht, pp. 9-28.

Dimou, P. (1994), 'The industrial district: a stage of a diffuse industrialization process - a case of Roanne', *European Planning* Studies, vol.1, no.2, pp. 23-38.

Dosi, G. (1988), 'The nature of the innovative process', in Dosi, G. et al (eds), *Technical change and economic theory,* Pinter Publishers, London, pp. 221-38.

Felsenstein, D. (1994), Book review essay (on Massey, D. et al, *High Tech Fantasies*, London, 1992), *Economic Geography,* vol. 70, no. 1, pp. 72-75.

Garofoli, G. (1991a), 'The Italian model of spatial development in the 1970s and 1980s' in Benko, G. and Dunford, M. (eds), *Industrial Change and Regional Development,* Belhaven Press, London, pp. 85-101.

Garofoli, G. (1991b), 'Local networks, innovation and policy in Italian industrial districts', in Bergman, E.M. et al (eds), *Regions Reconsidered*, Mansell, London, pp. 119-40.

Giddens, A, (1984), *The Constitution of Society: Outline of the Theory of Structuration,* Polity Press, Cambridge.

Glasmeier, A. (1994), 'Flexible districts, flexible regions? The institutional and cultural limits to districts in an era of globalization and technological paradigm shifts', in Amin, A. and Thrift N. (eds), *Globalization, Institutions, and Regional Development in Europe,* Oxford University Press, Oxford, pp. 118-46.

Grabher, G. (1993), 'Rediscovering the social in the economics of inter-firm relations', in Grabher, G. (ed.), *The Embedded Firm. On the Socioeconomics of Industrial Networks,* Routledge, London, pp. 1-31.

Granovetter, M. (1973), 'The strength of weak ties', *American Journal of Sociology,* vol. 78, no. 6, pp. 1360-80.

172

Granovetter, M. (1985), 'Economic action and social structure: the problem of embeddedness', *American Journal of Sociology,* vol. 91, no. 3, pp. 481-510.

Haraldsen, T. (1994), Teknologi, økonomi og rom - en teoretisk analyse av relasjoner mellom industrielle og territorielle endringsprosesser, doctoral dissertation, Department of Social and Economic Geography, Lund University, Lund University Press, Lund.

Haraldsen, T. (1995), 'Spatial conquest - the territorial extension of production systems', paper presented at the Regional Studies Association conference on "Regional Futures: Past and Present, East and West", Gothenburg, May.

Harrison, B. (1991), 'Industrial districts: old wine in new bottles?' Working Paper 90-35, School of Urban and Public Affairs, Carnegie-Mellon University.

Henry, N. et al (1995), 'Along the road: R&D, society and space. *Research Policy,* vol. 24, pp. 707-726.

Herrigel, G. (1996), 'Crisis in German decentralized production: unexpected rigidity and the challenge of an alternative form of flexible organization in Baden Württemberg', *European Urban and Regional Studies,* vol. 3, no. 1, pp. 33-52.

Hirst, P. and Zeitlin, J. (1990), *Flexible Specialisation vs. Post-Fordism: Theory, Evidence and Policy Implication,* Discussion paper for the workshop "Flexible Specialisation in Europe", Zürich, Switzerland, 25-26 October.

Håkansson, H. (1992), *Corporate Technological Behaviour. Co-operation and Networks,* Routledge, London.

Johnson, B. and Lundvall, B-Å. (1991), 'Flexibility and institutional learning' in Jessop, B. et al (eds), *The Politics of Flexibility. Restructuring State and Industry in Britain, Germany and Scandinavia,* Edward Elgar, Aldershot, pp. 33-49.

Lazonick, W. (1993), 'Industry cluster versus global webs: organizational capabilities in the American economy', *Industrial and Corporate Change,* vol. 2, pp. 1-24.

Lazonick, W. and Smith K. (1995), 'Organisation, innovation and competitive advantage: long-term investment and public policy', Project proposal, STEP Group, Oslo.

Leborgne, D. and Lipietz, A. (1988), 'New technologies, new modes of regulation: some spatial implications', *Environment and Planning D: Society and Space,* vol. 6, no. 3, pp. 263-80.

Leborgne, D. and Lipietz A. (1992), 'Conceptual fallacies and open questions on post-Fordism', in Storper, M. and Scott, A.J. (eds), *Pathways to Industrialization and Regional Development.* Routledge, London, pp. 332-48.

Lipietz, A. (1993), 'The local and the global: regional individuality of interregionalism?', *Transactions of the Institute of British Geographers, New Series,* vol. 18, 8-18.

Lipparini, A. and Lorenzoni, G. (1994), 'Strategic sourcing and organisational boundaries adjustment : a process-based perspective', paper presented at the workshop on "The Changing Boundaries of the Firm", European Management and Organisations in Transition (EMOT), European Science Foundation, Como, October.

Lipparini, A. and Sobrero, M. (1994), 'The glue and the pieces: entrepreneurship and innovation in small-firm networks'. *Journal of Business Venturing,* vol. 9, pp. 125-40.

Lorenz, E. (1992), 'Trust, community and cooperation: toward a theory of industrial districts', in Storper, M. and Scott, A.J. (eds), *Pathways to Industrialization and Regional Development.* Routledge, London, pp. 195-204.

Lundvall, B.-Å. (1992), 'Introduction' in Lundvall, B.-Å. (ed.), *National Systems of Innovation,* Pinter Publishers, London, pp. 1-19.

Lundvall, B.-Å. (1993), 'Explaining inter-firm cooperation and innovation: limits of the transaction-cost approach', in Grabher, G. (ed.), *The Embedded Firm. On the Socioeconomics of Industrial Networks,* Routledge, London, pp. 52-64.

Lundvall, B.-Å. and Johnson, B. (1994) 'The learning economy', *Journal of Industry Studies,* vol. 1, no.2, pp. 23-42.

Lyberaki, A. and Pesmazoglou, V. (1994), 'Mirages and miracles of European small and medium enterprise development', *European Planning Studies,* vol. 2, no. 4, pp. 499-521.

Malerba, F. (1993), 'The national system of innovation: Italy', in Nelson, R. (ed.), *National Innovation Systems. A Comparative Analysis,* Oxford University Press, New York and Oxford, pp. 230-259.

Marshall, A. (1919), *Industry and Trade*, Macmillan, London.

Martinelli, F. and Schoenberger E. (1991), 'Oligopoly is alive and well: notes for a broader discussion of flexible accumulation', in Benko, G. and Dunford, M. (eds) *Industrial Change and Regional Development,* Belhaven Press, London, pp. 117-132.

Morgan, K. (1995), 'Institutions, innovation and regional renewal. The development agency as animateur', paper presented at the Regional Studies Association conference on "Regional Futures: Past and Present, East and West", Gothenburg, May.

Patchell, J. (1993) 'From production systems to learning systems: lessons from Japan', Environment and Planning A, vol. 25, pp. 797-815.

Pedler, M. et al (1991), *The Learning Company.* McGraw-Hill, London.

Perroux, F. (1970), 'Note on the concept of "growth poles"', in McKee, Dean and Leahy (eds.) *Regional Economics: Theory and Practice,* The Free Press, New York, pp. 93-103.

Perulli, P. (1993), 'Towards a regionalization of industrial relations', *International Journal of Urban and Regional Research,* vol. 17, no. 1, pp. 98-113.

174

Piore, M. and Sabel, C. (1984), *The Second Industrial Divide: Possibilities for Prosperity,* Basic Books, New York.

Porter, M. (1990), *The Competitive Advantage of Nations.* Macmillan, London.

Pyke, F. (1994), *Small Firms, Technical Services and Inter-firm Cooperation.* International Institute for Labour Studies, Geneva.

Pyke, F. and Sengenberger, W. (1990), 'Introduction' in Pyke, F. Becattini, G. and Sengenberger, W. (eds), *Industrial Districts and Inter-firm Co-operation in Italy,* International Institute for Labour Studies, Geneva, pp. 1-9.

Reich, R. (1991), *The Work of Nations: Preparing Ourselves for 21st-Century Capitalism,* Knopf, New York.

Russo, M. (1989), 'Technical change and the industrial district: The role of inter-firm relations in the growth and transformation of ceramic tile production in Italy', in Goodman, E. and Bamford, J. (eds), *Small Firms and Industrial Districts in Italy,* Routledge, London, pp. 198-222.

Sabel, C. (1989), 'Flexible specialisation and the re-emergence of regional economies', in Hirst, P. and Zeitlin, J. (eds), *Reversing Industrial Decline? Industrial Structure and Policy in Britain and Her Competitors,* Berg Publisher, Oxford, pp. 17-70.

Sabel, C. (1992), 'Studied trust: building new forms of co-operation in a volatile economy', in Pyke, F. and Sengenberger, W. (eds), *Industrial Districts and Local Economic Regeneration,* International Institute for Labour Studies, Geneva, pp. 215-50.

Sally, R. (1994), 'Multinational enterprises, political economy and institutional theory: domestic embeddedness in the context of internationalization', *Review of International Political Economy,* vol 1, no. 1, pp. 161-192.

Sayer, A. (1992), 'Radical geography and Marxist political economy: towards a reevaluation', *Progress in Human Geography,* vol. 16, no. 3, pp. 343-360.

Sayer, A. and Walker R. (1992), *The New Social Economy: Rethinking the Division of Labor,* Basil Blackwell, Boston, MA.

Scott, A. (1992), 'The role of large producers in industrial districts: a case study of high technology systems houses in Southern California', *Regional Studies,* vol. 26, no. 3, pp. 265-275.

Semlinger, K. (1993), Small firms and outsourcing as flexibility reservoirs of large firms', in Grabher, G. (ed.), *The Embedded Firm. On the Socioeconomics of Industrial Networks,* Routledge, London, pp. 161-78.

Sheard, P. (1983), 'Auto-production systems in Japan: organisational and locational features', *Australian Geographical Studies,* vol.21, pp. 49-68.

Smith, K. (1994), 'New directions in research and technology policy: Identifying the key issues', *STEP-report,* No. 1, The STEP-group, Oslo.

Storper, M. (1995a), 'Regional technology coalitions: an essential dimension of national technology policy', *Research Policy,* vol. 24, pp. 895-911.

Storper, M. (1995b), 'The resurgence of regional economies, ten years later: the region as a nexus of untraded interdependencies', *European Urban and Regional Studies,* vol. 2, no. 3, pp. 191-221.

175

Storper, M. (1995c), 'Territorial development in the global learning economy: the challenge to developing countries', *Review of International Political Economy,* vol. 2, no. 3, pp. 394-424.

Storper, M. and Harrison B. (1990), 'Flexibility, hierarchy and regional development: the changing structure of industrial production systems and their forces of governance in the 1990s', UCLA, MIT/Carnegie-Mellon University, Los Angeles, Cambridge/Pittsburgh.

Storper, M. and Walker R. (1989), *The Capitalist Imperative. Territory, Technology, and Industrial Growth,* Basil Blackwell, New York.

Tolomelli, C. (1990), 'Policies to support innovation in Emilia-Romagna: experiences, prospects and theoretical aspects', in Alderman, N., Ciciotti, E. and Thwaites, A. (eds), *Technological Change in a Spatial Context: Theory, Empirical Evidence and Policy,* Springer-Verlag, Berlin, pp. 356-78.

Tödtling, F. (1995) *Firm Strategies and Restructuring in Globalized Economy,* IIR-Discussion 53, Institute for Urban and Regional Studies, Wirtschaftsuniversität, Vienna.

Varaldo, R. and Ferrucci, L. (1996), 'The evolutionary nature of the firm within industrial districts', *European Planning Studies,* vol. 4, no. 1, 27-34.

Weinstein, O. (1992), 'High technology and flexibility, in Cooke, P. et al (eds), *Towards Global Localisation,* UCL Press, London.

You, J.-I. and Wilkinson, F. (1994), 'Competition and co-operation: toward understanding industrial districts, *Review of Political Economy,* vol. 6, no. 3, pp. 259-278.

Zeitlin, J. (1992), 'Industrial districts and local economic regeneration: overview and comment', in Pyke, F. and Sengenberger W. (eds), *Industrial Districts and Local Economic Regeneration.* International Institute for Labour Studies, Geneva, pp. 279-94.

8 Globalisation and territorial economy: what future for the north of France as an old industrial area?

Federico Cuñat and Bernadette Thomas

Introduction

The question posed in this chapter is, will there be a new role for the traditional industrial areas of north west Europe in a globalised world economy? In particular, will areas with strong spatial proximities which stem from highly concentrated industrial activity established in the past have access to the global economy and, if so, under what conditions? Since the industrial revolution of the 19th century, such places have developed dense networks of interconnections. These networks originated in production and they generated many economic institutions associated with production. The place specific institutions created in this way sustained subsequent periods of growth and converted them into territorial development through the continuous spatial diffusion of dynamic activities (Perroux, 1963; Boudeville, 1968; Lasuen, 1969).

In the context of economic globalisation, growth is currently occurring in spaces organised on the basis of flows and exchange (Castells, 1989). Such growth tends to concentrate on a global scale in a small number of cities. Does this mean that historic industrial areas such as those of north west Europe are redundant once and for all? There is evidence to suggest that this might not be the case and that in some places resources are now being generated locally in a general context of industrial crisis (Rallet, 1993; Favereau, 1989). Central to understanding these crisis driven changes, the very nature of the firm itself needs to be looked at very closely. Principally, should the firm be looked on as an institution or as an organisation? The literature suggests that both interpretations are valid. Detailed examination shows that theories of exchange can be linked through the firm to theories of production. Indeed, empirical studies of industrial processes reveal different territorial forms of interdependence and cooperation between firms[1]. Network analysis, for example, concentrates on non-commercial interdependencies, as well as on the effects of spatial proximity on interdependence between firms.

177

Recognising these organisational and institutional dimensions of economic coordination helps to put the regulatory function of the market into perspective, showing that its influence is not all pervasive. It also clarifies the limits of the notion of the contract as a mechanism for managing interaction. Indeed, aspects of interaction related to the interdependencies between firms, and in particular to proximity, cannot be incorporated into contracts and resolved in a conventional transaction cost manner. Interactions cannot be reduced to the opposites of 'organisation' and 'market', because '...they are not of a commercial but an industrial nature, as they specifically belong to a production process' (Ravix and Torre, 1991, p. 377). This means that different processes and procedures in production give rise to different types of coordination between firms. Thus the contract, designed to cope with questions of pure exchange, becomes a very restricted instrument for dealing with organisational questions related to industrial cooperation.

Against this background a micro-economic approach is adopted in this chapter which concentrates on the production process to explain organisational forms and the internal characteristics of firms, as well as their external relations. This is the direction in which analyses of localised industrial systems has developed. Research in sociology, for example, focuses on networks of social relations within an economy, together with communal forms of understanding within these networks. The emphasis in understanding new industrial agglomerations is no longer on growth factors but on social and institutional factors instead - non-market relations, trust, social consensus, the institutional support of local business, agencies and traditions which favour innovation, acquisition of skills and the circulation of ideas (Granovetter and Swedberg, 1992). These shifts of emphasis suggest reasons why established local centres of production still have a part to play in the context of global networks.[2] Their strong institutionalisation reinforces growth and recognises appropriately the 'local' dimension which cannot be reduced solely to economic factors of proximity. These centres of agglomeration are seen as centres of representation and innovation belonging to global chains of production. They qualify as centres of industrial excellence, capable of providing well established networks of linkages, knowledge structures and institutions to a large collectivity of entrepreneurs.

The reorganisation of the textile sector in the North of France offers the opportunity to explore these ideas against the background of economic decline and industrial restructuring. An essential aspect of the analysis is the role played by spatial proximity in the search for new opportunities in entrepreneurial environments. Strong emphasis is put on social relations as a mechanism to produce coherence based on personal identification. What emerges from the analysis is a picture of strengthening social relations binding together an emerging small firms structure generated by processes of deconcentration among large textile firms that have had to attempt to cope with intensifying global competition. The discussion of the chapter is developed in three broadly chronological stages. First, the transformation of the Fordist structure that still dominated in the 1970s is examined as jobbing production, subcontracting and mail order distribution emerged in the 1980s. Second, the emergence in 1994 of the 'Cité de l'initiative' (textile park) as a socialised structure of interdependent production enterprises is examined as a mechanism to cope with the inequities of subcontracting. Third, the charter signed in 1995 by several large northern distributors is examined in detail as a template for the socialisation of interdependencies and interactions between both large business organisations and small firms in this old textile region. It is suggested that the creation of a charter between producers and distributors rests upon their experience of an economy based on solidarity, with proactive initiatives substituting for crisis management.

The redistribution of roles in the textile industry

Until the 1970s, the value chain of the textile industry in northern France was dominated by large, Fordist enterprises that embraced production and specialist services including marketing, finance and design. Figure 8.1 shows schematically the relationships involved in this system of production which characteristically was integrated (a filière), focused on large family firms, and hierarchical. The crisis in Fordism that emerged in the 1970s significantly impacted on this system through:

- the re-imposition of the discipline of markets rather than of hierarchies;
- internationalisation and globalisation of production and services;
- the creation of new spatial divisions of labour;
- the quickening of technological change;
- the emergence of new sectors of production and services;
- the generation of new methods of enterprise organisation and control together with new inter-organisational relationships; and
- the commodification of money and the creation of new financial instruments (Ekinsmyth et al, 1995).

Major restructuring was the immediate consequence. Large enterprises were faced with reduced incomes and the need for sophisticated inventory management and more paid-up capital. Many firms closed down.

However, the displacement of firms during the 1980s took place alongside the redistribution of roles within the textile industry. Buyer organisations rapidly incorporated the design function into their activities when they realised that it contributed an important part to overall value added. Trimming production seemed a viable solution to cope with uneven sales, and faced with substantial falls in demand, firms became cautious and produced only the amount they were sure to sell. Peaks were managed by transferring production to subcontractors both in France and abroad.

What has developed is a system of 'jobbing production', a model of subcontracting which works without stocks and with short supply periods. The subcontractors are supplied with everything and have simply to produce the garments. The system tends to reduce the time between the start of production and the moment when the product is sold to the client. It also represents a type of

179

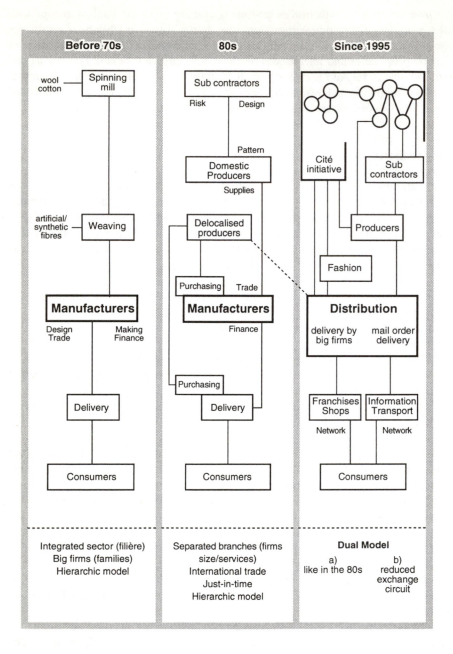

Before 70s	80s	Since 1995

Before 70s

wool cotton — Spinning mill

artificial/synthetic fibres — Weaving

Manufacturers
Design / Trade   Making / Finance

Delivery

Consumers

80s

Sub contractors
Risk   Design

Pattern
Domestic Producers
Supplies

Delocalised producers

Purchasing   Trade
Manufacturers
Finance

Purchasing
Delivery

Consumers

Since 1995

Cité initiative   Sub contractors

Producers

Fashion

Distribution
delivery by big firms   mail order delivery

Franchises Shops | Information Transport
Network   Network

Consumers

Integrated sector (filière)
Big firms (families)
Hierarchic model

Separated branches (firms size/services)
International trade
Just-in-time
Hierarchic model

Dual Model
a)   b)
like in the 80s   reduced exchange circuit

Figure 8.1 Evolution of the textile industry in the north of France

180

production which is capable of adapting to clients' quality and quantity requirements in a very short time. It is a system of total flexibility in which production is not standardised and there is a continuous search for adaptation.

Through jobbing production, the clothing producers shift the risks and fluctuations of global competition onto subcontractors. It is the subcontractors who have to manage fluctuations in demand and cope directly with the repercussions on their production. Large suppliers realised the advantages they could gain from dealing directly with subcontractors and this led to the creation of new functions in the textile production chain. The subcontracting garment manufacturers produced the patterns. They received fabric from the large suppliers but they had to supply all the other production inputs themselves. In turn, these firms had greater freedom to respond directly to the requests of distributors or designers.

This fragmentation and redistribution of roles has introduced radical changes into the textile industry. In particular, large clothing manufacturers with 400 to 800 employees have disappeared. However, besides these closures, many very small enterprises have emerged, thereby reducing the average size of textile firms. The employees are mainly female and, during the last few years technological innovation has changed the nature of their work. New demands for quality have resulted in further changes. Jobs have evolved. Manufacturing and unskilled tasks have tended to disappear and to be replaced by new tasks related to the maintenance of automated production processes. As a result, while the textile sector is declining generally, dynamic local situations have emerged, akin to those of Le Sentier in Paris[3], which have generated new relations between suppliers and subcontractors. Indeed, jobbing production has often been put forward as a 'miracle' solution capable of saving the textile sector (Lazzarato et al, 1993).

The modified production chain that merged in the 1980s is demonstrated diagrammatically in Figure 8.1. Dis-integration led to the functions of the filiere being separated, thus creating an hierarchic subcontracting system with just-in-time characteristics and increased international trade. The large French clothing manufacturers were, in effect, transformed into distributors employing French and foreign subcontractors to produce clothing lines to supply franchises and retailers, and for sale through mail order. The system put pressure on French subcontractors to reduce overheads and this they did by creating operations with fewer than 10-15 employees which, though on the margin of viability, were sufficiently small to escape trade union constraints.

Trois Suisses and Camaïeu are examples of large French clothing manufacturers that have operated in this way since the mid 1980s. Their subcontractors are small, responsive and flexible. When there is no work, the employees stay at home and receive a wage which corresponds to a 39 hours working week, regardless of the number of hours they have worked. The difference between their pay and the effective hours they have worked are verified twice per year and excess working hours are remunerated accordingly.

181

The purchasing departments of the large clothing distribution firms play a pivotal role in the operation of this clothing industry filiere. At Trois Suisses, buyers choose their collections and decide on quality with the aid of a central quality control service. At Redoute, in contrast, the nature of the collection is decided elsewhere in the firm, thus constraining the autonomy of buyers. The buyers themselves tend to have commercial rather than technical backgrounds, further affecting their purchasing practices. Paid through bonuses, their incomes depend on their own efficiency, not the technical capabilities of their subcontractors.

To combat the problems associated with operating in the textile industry of northern France in the mid 1990s, small and medium sized enterprises have in recent years created associations as a social and institutional response to their subordination and powerlessness within subcontractor systems (Figure 8.1). These place-specific associations are quite different to clusters generated through local industrialisation processes and the experience of places such as Le Sentier in Paris, for example. To explore the nature of these differences, the character and nature of the 'Cité' is examined in detail in the next section.

Territorial initiative : the 'Cité de l'initiative'

Based on the principles of autonomy and independence, the 'Cité de l'initiative' (a textile park) is an association that was founded in 1994 by small, more or less interdependent enterprises on a single site (Figure 8.1). The production units specialise in particular parts of the production process. They also provide services such as accountancy, technical support and training. A single location aims to facilitate cooperation, exchange and communication, a prerequisite of entrepreneurial dynamics. Coordination and division of labour depends on common interests and relations of solidarity between the different production units. Collective risk-taking reflects the dynamics of this association. What distinguishes this environment is that the participants share the same ethic. This corresponds to the creation of a particular externality, a collective good built up by the participants which they can mobilise for the benefit of their own organisations (Perrin, 1989).

The firms located in the 'Cité' are all connected with the textile industry. Their guiding principle is to design a product or a collection of products to client demand, and to manufacture and deliver it. All the skills of the textile chain are therefore present on the site:

- the designer (modeliste) produces the pattern (one company, SARI, specialises in fashion techniques);
- the time-share textile centre (CSTTP) carries out all the production operations for a given pattern - digitising the design, transposing it into different sizes, setting up the pattern, computerising it, devising a cutting plan which defines the use of the fabric, tracing and cutting it;

- a workshop which produces prototypes or a whole collection of designs is able to respond to the demand of any client who uses the other production units on the site. This may range from a single operation to the whole production process of clearly laid down tasks.

Three clothing enterprises provide examples of the types of small firm operating as part of the 'Cité'. IMC ('Impact Mode Creation') specialises in small production runs to fill gaps in existing garment collections, and supplies small quantities or single collections to shops. It is run by workers with a broad range of sophisticated skills. Flandres Ateliers has been running for three years and is located in a converted textile factory. It obtains state benefits for sheltered workplaces as compensation for the low productivity of its handicapped employees. One of its aims is to reintegrate disabled persons into a normal work environment. This workshop specialises in repetitive tasks, including some which may present technical difficulties. L'Atelier Decalonne comprises two organisations which practice a technical division of labour. The fashion industry produces large production runs which are nevertheless able to respond to demand thanks to 'just-in-time' techniques. The centralised cutting service (Decalonne Service) has high-tech automated cutting equipment which is driven by computer disks either supplied by the clients or produced by the Textile Service Centre of the 'Cité'. It serves a range of different clients including garment producers and clients ordering complete production. In this latter case, the atelier accepts the order and subcontracts work out to its network of small enterprises in the Cité.

An essential part of the 'Cité' are the collective services it provides for its members. These services include a nursery for employees, a canteen and business services, as well as management and administrative advice for small firms. For example, GIAP supplies management and accountancy (computerised work schedules, forecasts, budgets and accounts, and so on) which enable small firms to exercise total management control. Another firm (France-Tri) sells tagging and sorting equipment for garments of the kind that are also used in laundries. These machines have a bar code or other electronic devices to recognise and collate items from the same firm or from a designated individual. This firm installs and provides maintenance for these Italian machines, and trains the personnel to use them.

A key characteristic of the firms in the 'Cité' is their under-investment which, in turn, is a function of their size. In general, the smaller the enterprise the less it is capitalised. However, although these small firms invest little, they are able to maintain competitive prices. The 'Cité' provides a solution to this technological problem which would otherwise greatly affects the costs and performance of these small firms. The provision of a centralised cutting workshop is a viable alternative to the separate provision of the same facilities in every small firm. It enables all the firms on the site, including single small units, to share it and gain access to advanced technology without tying up their working capital.

The 'Cité' as a place of production has, therefore, a number of distinctive characteristics:

- it ensures flexibility of the work force among all the firms on the site;
- commercial contracts with clients are very flexible as the firms on the site can cope with demands at every stage of the production process;
- all firms seek to pay wages at market rates for all employees to harmonise the work force. Firms would refuse orders which would make such wage payments impossible;
- a qualified work force supplies high quality products and services;
- a mutual support structure between the production units increases capitalisation and raises the level of investment in advanced technology in connection with conventional work methods. Such integration of material and non-material (conception, design of collections, and so on) aspects of manufacture revalues traditional production;
- member firms develop functional and constructive solidarity.

This mode of production shortens the production process. It also constitutes a more complete method of modernising subcontracting as it integrates the upstream phases of the process. Due to its novelty, this process is difficult to asses. Also its innovative nature differs from known local industrialisation processes such as industrial clusters or the experience of Le Sentier. The main differences consist of its productive form of socialisation which emphasises integration into a value chain, its method of recruitment and its network structure underpinned and driven by solidarity. This type of solidarity does not correspond to the sociability found in the family or on certain estates, or to the solidarity constructed by immigrant communities. It is a solidarity derived from a mode of production which, unlike Le Sentier, is not entirely motivated by commercial logic.

The experience of the 'Cité' in Roubaix tends to act as a model for other geographic areas and sectors of production (e.g. the printing industry in Dunkerque or furniture production, for example). The institutionalisation of the 'Cité' also involves not only the public sector but also the private and the social sectors. Links between private and institutional agents are known to contribute to structural efficiency. Thus, the Roubaix experience could be conceived as a global phenomenon in terms of specific localised production. It may constitute a transferable model which responds to strictly economic needs. Nevertheless, social and political conditions have ensured its viability. Its origin lies in the closure of the Phildar factory and the strategy of this large firm (which formerly produced and marketed knitting yarn) to restructure itself towards the service sector. The employees broadened this initiative by developing a chain of complementary activities. To achieve this, they networked within an association which provides the technical assistance of specialists from the client sector.[4.] Thus, this form of alternative production in the 'Cité' builds on work experience within large firms.

Former employees of the textile industry have created a number of firms in the 'Cité' by starting up alternative forms of production in response to the restructuring crisis in the textile industry. Paradoxically, when large firms released direct control over their work force they only appear to have given up their authority. By subcontracting and making use of this new form of productive solidarity which has emerged from the crisis, those who commission work have developed a very flexible work method which suits their needs and constitutes their contribution to the restructuring of the labour market. Thus, this new entrepreneurship enjoys only relative autonomy. Subjected to the logic which dominates economic and social production, the new entrepreneurs depend on those who issue the orders, unless they associate with each other to increase their power in the market place. In reality, reorganisation is regulated by the norms of the large firm which has created this new entrepreneurship outside its ranks and guides its activity indirectly. Indeed, the competitive position of the new entrepreneurs is conditioned by two constraints. Price limits force them to squeeze internal production costs (comparative advantage), while the incorporation of non-material elements into the production system enhances access to value added and distinguishes this local production from Third World production (competitive advantage).

The 'Cité' initiative is sustained by financial, professional and moral solidarity which is based on social relations of mutual support. Purely commercial transactions constitute only a secondary aspect of these relations. Services are billed at market value, but more often direct services are provided as non monetary arrangements. At the same time, great rigour is attributed to management control. What has emerged here is a new employer group constituted by small enterprises. They endeavour to comply with a charter based on a code of conduct which they would like to impose outside the 'Cité'. One example is to refuse the fragmentation of orders which interrupts serial production and prevents economies of scale. Thus, the small firms of the 'Cité' act as pressure groups on those who commission orders, trying to regain power and at the same time limit the constraints placed upon them. Their ambition is to become a technical centre in the fashion and garment industry by incorporating the design function into production and by adopting new technologies to modernise subcontracting.

A textile and clothing charter in the North of France

The socialising of relationships and interactions in the textile filière of northern France is now going much further than the defensive subcontractor associations, such as the 'Cité' discussed in the previous section, to involve the large distribution firms. Representatives of several large northern distributors[5] and the employer organisations of the clothing and textile industry signed a charter in 1995 entitled *Commitment to Progress in the Production Chain: Textile - Clothing - Distribution*. According to this charter, the signatories intend to promote a new model of development. It should enable them to revitalise the regional textile chain of northern France which has been in deep crisis for many years. In order to reach this objective, they concentrate their efforts on changing the relations between the different participants in the chain.

The idea of the charter is to improve commercial performance by developing 'jobbing production' in conjunction with 'just-in-time' practices, thanks to preferential relations between the actors concerned and the support of the public sector.   Such a development requires cultural as well as structural changes. Cooperation, based on the exchange of information, coordination of the internal organisation of the participant firms, and a commonly designed system of evaluation, substitutes for calls for tenders which put producers into direct competition. Partners agree a minimum of common rules, recognised and accepted by all.

Contrary to traditional commercial exchange, whereby payment relieves firms from all contractual obligations, the charter introduces a commitment over time with the aim of providing stability and security of exchange. In the first instance, it guarantees cooperation despite divergent interests.   Certain practices are thus changed. For example, subcontractors could be tempted to reduce quality in order to curb costs, or distributors could delay payment to increase profit.[6]   The charter also enables the partners to dispense with continuous negotiations which increase transaction costs.[7]   It provides economies of knowledge and time by encouraging electronic document interchange (EDI), or reducing paper work and cumbersome administrative procedures. Codification of exchanges and development of routines make such exchange relations more viable than contracts which are always incomplete (Baudry, 1994; Eymard-Duvernay, 1994). Thus, the charter introduces a certain 'morality' into the exchanges, without preventing the various signatories from pursuing their respective interests. It differs from instrumental rationality which regulates traditional economic exchange. It seems to be rooted in social relations and customs (Granovetter, 1985).   It can be interpreted as a mechanism of social exchange between small and large firms, and between industrial and commercial firms in a competitive situation.

It is possible to interpret these changes as the simultaneous introduction of new mechanisms of behavioural coordination between each of the two categories of protagonists - the textile industrialists and the distributors - as well as within these categories themselves. Similarly, the agents have to honour the commitments within each firm of producers or distributors. Thus, the charter links all the partners to each other. It broadens commercial exchange, as it guarantees the reproduction of the relations between firms - besides individual implications - while assuring the exchange of goods, services and capital. Such enlarged procedures reduce standard economic exchange to a limited category of exchange (Homans, 1994; Saglio, 1991). Thus, the conditions of stable relations between the two groups differ from the conditions of equilibrium stemming from standard mechanisms of price formation in the market place.

Besides strengthening links between independent productive organisations, the Charter also aims to regulate the conditions which balance social structures between different social groups at the local level. While the identity of the signatories of such agreements reinforces social proximities due to spatial proximity, other aspects contribute as well. This is confirmed by the importance of respecting procedures instead of leaving price formation essentially to market forces. Therefore, exchange loses its characteristic of spontaneity, as implied in a purely instrumental representation of coordination. Indeed, exchanges undergo time lags which stem from the terms of the Charter. Distributors relinquish their market powers at a particular moment in time to reap non-quantifiable advantages at a later stage. Similarly, producers accept exchange conditions (low prices, flexibility, tighter deadlines etc.) in the hope of engaging in future exchanges with distributors which will yield as yet unknown benefits. These expectations do not concern bilateral exchanges between individuals. They apply to whole groups of agents (URIC,[8], GRITT[9] or distributors). Saglio (1991) underlines the importance of this issue while referring to Peter Blau's (1964) notion of the temporality of exchange. The latter has a broader understanding of exchange in comparison with purely economic exchange, as the counterpart in the initial position is not yet specified at the time when the exchange is taking place. It is clear that in the absence of financial relations which create hierarchical integration, these conditions can only be accepted if social codes are shared territorially. Thus, it is indispensable to share territorial resources to sustain the viability of such agreements.

The Charter as a system of obligations

The system of exchange regulation that is incorporated into the Charter differs from contracts. According to anthropological methods of analysis,[10] it could be construed as a system of obligations which establish time lagged links of equality, subordination and reciprocity. Such a system of exchange has several aspects of which three are outlined here; collectivity, non-utilitarianism and regulation by law.[11]

Collective exchange

Individual signatories of the Charter are not signing as private individuals but as legal persons: the firm, the trade union and so on.[12] Moreover, these collectivities cannot be reduced to simple interest groups. They are not only motivated by common interest but also by 'kinship' relations. Without taking the notion of 'kinship' literally, it may apply to employers who operate in a family mode with lineages and branches of descendants or dissidents. Many leaders of distribution firms who have signed such charters belong to the 'Mulliez' [13] culture. They all have strong regional roots.

An essentially non utilitarian system of exchange

The Charter can be conceived of as a system of mutual services, albeit without the exchange of symbolic goods as practised by primitive societies. However, the highly non-material content of the Charter or a common ethic may substitute for symbolic goods. In any case, economic aspects are compounded with other social and political dimensions, which may be contained in the notion of the 'total social fact' ('fait social total').[14] The Charter is deeply rooted in its context. It puts territorial identity before economic implications; it motivates the participants to commit themselves to the future of a territory to which they belong. It is linked to a system of allegiances whose content depends directly on the local history from which it stems, as well as on the conditions within which it operates.

An exchange based on the rule of law

Three key rules of law regulate the system of primitive reciprocity: the obligation to give, the obligation to receive and the obligation to give back. The Charter determines the rules of exchange. It creates, therefore, a moral obligation to comply. Those who fail to comply are excluding themselves from the group of leaders who innovate and induce change. Obligations are not defined as clearly as in ordinary contracts. Implicit social practices vest commitment in the trust which is attributed to partners. However, such practices only become apparent when the power relations between the different partners are being breached. Recently, for example, the distributors have been joining up with powerful purchasing agencies and are now dominating the production chain. The Charter aims at more 'equitable' relations between the various stages of the production chain by mobilising the flexibility which is inherent in territorial organisation. Complementary relations are extended over time, which modifies the rules of competition. The 'zero-sum' game is replaced by a 'win-win' game, including the consumer who benefits from 'just-in-time' products at a favourable price.[15] The role of the Charter in social production is not yet clear, considering the novelty of this experience. Even if it simply triggers cooperative behaviour, this type of industrial relation could be usefully extended to create systems of quasi-integration. However, local subcontractors may be pressured by distributors to become increasingly flexible. Regardless of these options, firms assume a role of regulator which they did not have under 'Fordism' (Coriat and Weinstein, 1995). This new role plays an essential part. It represents a collective will to regulate professional relations and to control society by combating the grey economy which is very widespread and has been encouraged indirectly by certain participants until now.[16] The proper functioning of the new system is thus in the interest of all types of organisations, including those which deal with social peace.

A new labour convention

Practices incorporated into relations between firms are also reflected in labour relations. They give priority to consultations and negotiation. Traditional labour relations, based on the Fordist-Taylorist model which establishes divisions within the firm, are replaced by new relations aimed at gradually encompassing all employees. While participatory management techniques are already well developed within certain firms (for example, Auchan and Promod), they are not known by the majority of textile and clothing enterprises which are still very influenced by models of hierarchical control. The expected advantages of participatory practices do not only concern a specific partner but all the signatories, as well as others who will adhere to the Charter. By determining the boundaries of a social group, the Charter clearly belongs to a specific group of participants who wish to distinguish themselves from those who do not adhere to its ethic. The participating distributors want to be recognised as a group whose practices differ from prior practices adopted by whoever took part in this market. GRIT and URIC, which represent the textile and clothing industries respectively, want to broaden their sphere of influence.[17] The Charter contributes concretely to the construction of a collective identity by bringing together participants who intend to be actively involved in steering the local economy rather that to become the victims of the crisis. The fact that the professional body of employers (Maison des Professions[18]) has signed the Charter is very significant. It means that the employers want to be seen to renew themselves in the urban market place.

Questions not conclusions

What has been demonstrated in this chapter is that the crisis in Fordism, and in particular the internationalisation and globalisation of production, has stimulated a range of responses in the production chain of the textile industry of northern France. Fordist production initially gave way to a subcontract system dominated by large manufacturers and distribution firms. What has developed subsequently, and especially in the 1990s, are increasingly socialised relationships between firms substituting trust, cooperation and an obligation system for contractual market exchange. The first socialisation appeared in defensive alliances between subcontractors to combat the power and domination of the large distribution firms. It has now been extended much further through the Charter discussed here to embrace large distribution firms, employer organisations and subcontractors.

The question that now arises is, will the current process of socialisation examined in detail in this chapter result in a new industrial model? Is an alternative emerging within the productive sector and its production units? Is the process which has been triggered by the large textile manufacturers, and reorganised according to norms which they influence, able to evolve into forms of coordination which reinforce the identification of each of the principal groups involved? If the objective is to generate a collective dynamic within the textile industry with an internal as well as an external identity, an intermediary collective learning structure would need to be set up. As the concept of 'pure economic exchange' is too limited to analyse all the relations which are linking these firms in northern France, their current approach could be construed as an intermediary practice of regulation. The reason is that, besides the pure market and the state, the social and physical proximity between the actors concerned is a factor which produces trust. It is in this context that this type of regulation could take shape, provided there exists an intermediary group which represents the general interest. The recently signed charter between the textile firms in northern France may point in that direction.

Notes

1.   For further discussion see our 1995 study 'Communication and publicité, une filière métropolitaine', in Savy, M. and Veltz, P. (eds), *Economie Global et Reinvention du Local*, Datar-Ed, de l'Aube.

2.   See Amin and Thrift (1992) on London and the industrial districts in Tuscany (p. 416).

3.   'Le Sentier' is a geographic area in the centre of Paris with a high concentration of garment industries. By extension, 'Le Sentier' has come to mean a flexible system of production totally dominated by the market which relies on short chain production. 'Le Sentier' has evolved under special conditions; inner city location, strong solidarity between immigrant communities (Jews, Asians), high creativity, loose compliance with the law. For a more complete analysis see Lazzarato et al, 1993.

4.   This is the case for the mail order sector which has resorted to jobbing production and just-in-time production to supply items for its existing collections and for their intermediary catalogues.

5.   They include the following firms: Auchan, Camaïeu, Le Redoute, Les Trois Suisse, Kiabi and Diramonde.

6.   Cf §6 of the Charter which stipulates that '... production is committed to quality throughout all the stages of the chain. This approach is based on clearly defined performance specifications and responsibilities for each phase of the production process, together with the certification of production processes, in order to avoid redundant and costly controls.'

7.   Cf §7 of the Charter '... partners finding themselves in the same commercial relation agree on a rational and homogeneous system of determining costs (value analysis) under predetermined production conditions.'

8.   URIC: Union Régionale des Industries de Confection (regional union of the garment industry).

9.   GRIT: Groupement Régionale des Industries Textiles (regional association of textile industries)

10.   See Mauss' (1973) essay on 'the gift'.

11. A collective, mainly non-utilitarian exchange incorporated in a mandatory legal system.

12. '... it is not individuals but collectives which engage in mutual obligations, exchanges and contracts.' (Mauss, 1973)

13. Originally active in textile production (Phildar), the Mulliez family has become the core of an association of major distributors (Auchan, Decathlon, Flunch, Boulanger, Leroy-Merlin). Moreover, they held 45 per cent of Trois Suisse International shares. This family has initiated human resource management with very active participation.

14. 'Fait social total' is a concept of integrative sociology coined by Mauss.

15. Cf §8 of the Charter, '... in order to ensure adequate profits in the long term for all concerned.'

16. Cf §10 of the Charter, '... the producers and the distributors commit themselves to respect, and request all others to honour all legal dispositions concerning labour legislation and the allocation of social contributions.'

17. 'the firms which will benefit from their participation in the Charter.'

18. The 'Maison des Professions' is the main employer organisation in France in terms of members and turnover. Founded in 1936 on a territorial not a sectoral basis, it includes at present some 1500 enterprises in Nord - Pas-de-Calais. Eighty per cent of these enterprises are located in the Lille agglomeration and represent the proactive membership which generates ideas on modernising firms.

References

Amin, A. and Thrift, N. (1992), 'Neo-Marshallian nodes in global networks', *International Journal of Urban and Regional Research,* vol. 16, pp. 571-587.

Baudry, B. (1994), 'De la confiance dans la relation d'emploi et de sous-traitance', *Sociologie de Travail,* vol. 1, pp. 43-59.

Blau, P.M. (1964), *Exchange and Power in Social Life,* John Wiley , New York.

Boudeville, J.R. (1963), *L'Espace et Les Pôles de Croissance,* PUF, Paris.

Castells, M. (1989), *The Informational City. Information Technology, Economic Restructuring and the Urban-Regional Process,* Blackwell, Oxford.

Coriat, B. and Weinstein, O. (1995), *Les Nouvelles Théories de l'Entreprise,* Le Livre de Poche, Série Références.

Ekinsmyth, C., Hallsworth, A.G., Leonard, S. and Taylor M. (1995), 'Stability and instability: the uncertainty of economic geography, *Area,* vol. 27, pp. 289-299.

Eymard-Duvernay, F. (1994), 'Coordination des échanges par l'entreprise et qualité des biens', in Orlean, A. (ed.), *Analyse Economique des Convennons,* PUF, Paris, pp. 307-334.

Favereau, O. (1986), 'Marchés internes, marchés externes. La formalisation du rôle des conventions dans l'allocation des ressources', in Salais, R. and Thevenot, L. (eds), *Le Travail. Marchés, Règles, Conventions,* Economica, Paris.

Favereau, O. (1989) 'Marchés internes, marchés externes', *Revue Economique, No. Spécial Economie des Conventions,* vol. 40, pp. 273-328.

Granovetter, M and Swedberg R. (eds) (1992), *The Sociology of Economic Life,* Westview Press, Bolder, CO.

Granovetter, M. (1985), 'Economic action and social structure: the problem of embeddedness', *American Journal of Sociology,* vol. 91, pp. 481-510.

Homans, G.T. (1958), 'Social behavior as exchange', *American Journal of Sociology,* vol. 63, pp. 597-607.

Lasuen, J.R. (1969), 'On growth poles', *Urban Studies, vol. 6,* pp. 137-161.

Lazarato, M., Moulier Doutang Y., Negri A, and Santilli G., (1993), *Des Entreprises Pas Comme Les Autres - Benetton en Italie - Le Sentier à Paris,* Publisud, Paris.

Mauss, M. (1973), 'Essai sur le don. Forme et raison de l'échange dans les sociétés archalques', in *Sociologie et Anthropologie,* PUF, Paris, pp. 145-284.

Orlean A. (ed.) (1994), *Analyse Economique des Conventions,* PUF, Paris.

Perrin, J-C. (1989), *Notes du C.E.R.,* vol. 104, Aix-en-Provenceé.

Perroux, F. (1963), 'Les industries motrices et la croissance d'une économie nationale',*Cahiers de l'I.S.E.A.,* vol. 2, pp. 151-180.

Rallet, A. (1993), 'Choix de proximité et processus d'innovation technologique', *Revue d'Economie Régionale et Urbaine,* No. Spécial 'Economie de proximités', vol. 6, pp. 365-386.

Ravix, J.L. and Torre, A. (1991), 'Eléments pour une analyse industrielle des Systémes localisés de production', *Revue d'Economie Régionale et Urbaine,* No. Special 'Milieux innovateurs: réseaux d'innovation', vol. 3/4, pp.375-390.

Saglio, J. (1991), 'Echange social et identité collective dans les systémes industriels', *Sociologie du Travail,* vol. 4, pp. 529-544.

9 Networks of small manufacturers in the USA: creating embeddedness

Edward J. Malecki and Deborah M. Tootle

Introduction

Networks are by now a well-known - if still incompletely understood - means by which firms interact with their environment. Building upon existing research on post-Fordism, flexible specialisation and regulation theory, Yeung (1994) emphasises that there are at least three types of network relations that firms must manage: intra-firm (especially within large enterprises), inter-firm, and extra-firm interactions with non-firm organisations such as governments. These provide a useful framework to which previous research has not always adhered, and they make us cognisant both of intra-firm channels of communication and of the fact that interactions with governmental and other institutions also need to be managed and understood. This chapter deals with the most conventional of Yeung's categories: inter-firm networks, specifically those that are formed in order to promote interaction where it did not exist prior to the establishment of the network. The networks studied comprise an American variant of the networks found in Europe.

The nature of networks

There are several streams of research that inform the study of inter-firm networks. Two of these - in business management and marketing, and in sociology and organization studies - are related only through the fact that their contributors are not well known to geographers. The best compilation of this work is the volume edited by Nohria and Eccles (1992) in which two contributions are of particular interest. Kanter and Eccles (1992) distinguish between academic usage of the word 'network,' which tends to focus on their characteristics and consequences, and the usage of practitioners, who uses 'network' as a verb to describe networking activities. A second paper by Perrow (1992) emphasises that networks are only one form of economic organisation. Others include integrated firms, conglomerates, jont ventures and holding companies, and subcontracting networks of large corporations (see also Storper and Harrison, 1991). What must be kept in mind is the essentially human character of business networks, which begin and remain as fundamentally inter-personal contacts (Aldrich and Zimmer, 1986; Dubini and Aldrich, 1991; Håkansson, 1989; Illeris and Jakobsen, 1990).

Somewhat related are two topics which have received attention from geographers recently: *strategic alliances*, of (mostly) large corporations (Ahern, 1993a, 1993b; Anderson, 1995; Dicken, 1992; Eriksson, 1995), and *industrial districts*, which are commonly described in terms of the 'networks' of firms present (Cooke and Morgan, 1993). The common American theme here is the Silicon Valley phenomenon, which some believe can be created elsewhere if the right conditions and values are created (Saxenian, 1994). Perrow (1992, pp. 456-457) believes that, within the USA, only in Silicon Valley are small-firm networks found that approach the model of industrial districts in Europe and Japan.

A final literature that bears directly on this chapter concerns the issues of proximity, local culture and interaction. Gertler (1993; 1995) emphasises the interaction between producers of machinery and new users in order effectively to implement and assimilate new technology embodied in complex machinery. The inter-firm interaction he describes is found in numerous other settings, including technology districts (Storper, 1993) and new-firm formation (Sweeney, 1991). This research describes environments where firms take advantage of agglomeration and proximity to utilise nearby sources of information, skilled labour, technology, and capital. However, the connection between agglomeration and flexibility has been overemphasised, since agglomeration does not ensure flexible specialisation, and flexibility is present in non-agglomerated settings (Malecki, 1995; Phelps, 1992).

In contrast to success stories of localised networks, now well-worn from nearly a decade of study, are places where networks have not developed, where innovation and technology are not native to the local culture and economy, and where firms struggle to remain competitive. This is the setting MacPherson has described so well in several studies on the Buffalo region (MacPherson, 1991, 1992). Other studies of non-networked regions include North Florida (Malecki and Veldhoen, 1993) and Norway (Vatne, 1995). In the absence of local networks, small firms can become dependent on a single large customer, as Young et al (1994) have shown. In general, however, firms in circumstances that fail to provide a diverse array of information and other resources in the local environment turn to external sources. In effect, they substitute non-local resources for the sparse (or non-existent) set of local resources (Vaessen and Keeble, 1993).

The body of research built in the context of industrial networks sheds additional light on manufacturing networks (Axelsson and Easton, 1992; Håkansson, 1989). Networks are aggregations of relationships among firms, which are distinct from short-term interactions (Easton, 1992, p. 8). These relationships comprise four elements:

- mutual orientation;
- dependence upon each other;
- bonds of various kinds and strengths; and
- the investments each firm has made in the relationship.

Looked at in this way, an important characteristic results from the bonds between firms. We would expect network structures to be somewhat stable, but also to change gradually in response to changes external and internal to the network.

Industrial network research tends to address network relationships which apply generally rather than only to small firms. For firms large and small, network relationships (linkages) permit access to resources not found internally. However, these external resources are especially essential for small firms, and it is for small firms that cooperation is most important (Mariti, 1990; Robertson and Langlois, 1995).

Operating within networks are two dialectical processes; competition and cooperation. Together, these processes create relationships that are stable but not static (Håkansson, 1992).

> The continuing processes of interaction between firms are stabilised since they take place within the context of existing relationships. However such relationships are also changing, partly in response to events external to the relationships and partly because of the transactions which help to define them. . . Network inertia and interdependencies slow and shape change. Thus networks do not have life cycles. They transform over time, merge, shift in focus and membership. . . The continuous interaction between firms offers, on the one hand, the opportunity for innovation and, on the other, the existence of a known and predictable environment in which it can be realised. (Easton, 1992, pp. 23-24)

A great deal of what flows through networks - on which network relationships depend - is information or knowledge. Thus, geographers have distinguished between *linkages* involving physical transactions and *contact networks* which involved intangible interchange (Thorngren, 1970). Gelsing (1992) distinguishes between a *trade network*, involving linkages between producers and users of traded goods and services, and a *knowledge network*, where the focus is on the flow of information and exchange of knowledge irrespective of its connection to the flow of goods. Linkage studies have a substantial history in economic geography (Hoare, 1985), and it has proven somewhat easier to account for physical flows based on economic exchange than relationships involving coordination and communication where goods are not exchanged (Easton and Araujo, 1992). More recent, understanding of how firms operate shows clearly that little production can take place without a great deal of knowledge from many sources.

Information networks: gatekeepers and informal knowledge networks

The importance to firm vitality of information from external sources is well established (Leonard-Barton, 1995; Lewis, 1995; Nonaka and Takeuchi, 1995). Obvious sources that were once considered mainly, or only, for their material links include suppliers and customers - relations formed and maintained through conventional production-related input-output relations. Suppliers and customers continue to be among the most important sources of knowledge for firms, and are the key actors in industrial networks (Axelsson and Easton, 1992).

Gatekeepers are key individuals in knowledge networks. They serve as a bridge between organisations, frequently translating across discipline-specific terminologies and organisational cultures. They are proactive in acquiring external information and, although they might use it personally, they also are keenly interested in passing it on to others in the organisation for their use (Falemo, 1989; Macdonald and Williams, 1994). Their role in selecting and filtering information is critical, but the role is rarely institutionalised or formalised. Most importantly, gatekeepers see their acquisition of information as part of a *quid pro quo*, whereby they are obligated to supply information in return, as a means of building or maintaining trust in their counterparts in other organisations. Therefore, despite the fact that informal information is the most common and often the most important means of interfirm information flow, it is channeled through gatekeepers who may retain it for their own use rather than to pass it on to others in the organisation (Macdonald and Williams, 1993).

An especially important group of gatekeepers for local development are 'community entrepreneurs' or 'social entrepreneurs' who have the development of the local community as a goal (Cromie et al, 1993; Johannisson, 1990; Johannisson and Nilsson, 1989). These 'key individuals' use their extensive personal contacts to communicate across sectors (Stöhr, 1990). They can provide the necessary sorting and evaluating that others - especially small firm owners and managers - are less able to do (Rosenfeld, 1992, p. 315). Lorenz (1992) refers to these as 'political entrepreneurs'- policymakers who encourage inter-firm communication and cooperation.

Sociological contributions: trust and social capital

Trust

The place of networks in understanding economic relations requires us to understand the nature of a social, rather than an economic variable: *trust*. Arms-length contractual relations, which Granovetter (1985) considers as functional substitutes for trust, are typically legalistic and may thwart the development of cooperative relationships. Sako (1992, pp. 37-47) details three types of trust. The first, contractual trust, is essentially the mutual expectation that promises made are kept.

The second type, competence trust, concerns technical and managerial competence to carry out a task, and is demonstrated in accepting goods from a supplier without inspecting them. The third type, goodwill trust, refers to mutual expectations of open commitment to each other, seen in a willingness to do more than is formally expected, such as sharing of information. Sabel (1992) uses 'negotiated loyalty' to refer to goodwill trust.

Trust creates and reinforces mutual obligations and cooperation, as opposed to legalistic, arms-length contractual relations (Casson, 1990, pp.105-124; Sako, 1992). Trust requires personal relationships that transcend the contact at hand, reinforced by face-to-face relationships. Trust relationships are typically informal and may result in a supplier exceeding the contractual requirements, whether by early delivery, higher quality, or some other means of assuring goodwill (Sako, 1992). The legalistic nature of relations among Anglo-Saxon firms (arm's-length contractual relations, or ACR) contrasts sharply with the obligational contractual relations (OCR) found among Japanese firms at the opposite end of the spectrum of possible trading relationships. OCR is not only an economic relationship, but also a social relation between trading partners based on mutual trust. Because of this underpinning, transactions can take place without prior agreement or specification of all the terms and conditions of trade. Casson (1990) suggests that Americans' reliance on arms-length, legal transactions may thwart the cooperative relationships found in other countries. In a regional setting, especially in innovative milieux, trust operates as the short-cut mechanism for communication and cooperation between firms (Hansen, 1992). Trust and the embeddedness of economic relationships into 'the deeper social fabric,' or the communal, non-economic institutions of the local area, is a key part of what distinguishes industrial districts from other localised agglomerations of firms (Harrison, 1992, p. 479).

Trust is not created without effort. The investment made in order to build trust, measured in people and their time, is a 'soft investment,' but represents a significant investment of resources (Easton, 1992, pp.13-14).

> The basis for trust seems to be that the actors have certain important values and norms in common. . . Through the combinations of economic and social relationships in the network, the information becomes rich, redundant and cheap. (Hertz, 1992, p. 110)

Among firms, trust permits informal, mutually obligational relationships to function, in contrast to formal, arms length relationships. One of the major functions of networks-as-policy is to encourage trust-based interactions among firms which have had little previous contact, or contact only as competitors.

Trust is difficult to measure, of course, but it forms the basis for the interactions that define a culture - national, regional, or local - and allow values and norms to be passed on to succeeding generations (Fukuyama, 1995). A single person cannot possess trust; it takes at least two to trust. Trust - or its absence - is an all-pervasive reality of the culture of a place (Lipnack and Stamps, 1995, p. 197). Cooperation may result from the presence of trust within communities or from a set of shared social norms; these are fragile and can break down from a number of causes (Lorenz, 1992).

Social capital

Social capital is 'a capability that arises from the prevalence of trust in a society or in certain parts of it' (Fukuyama, 1995, p. 26). Social capital is not created or acquired by a rational investment decision of an individual in education or training. It occurs through the acquisition of norms (or 'virtues' such as loyalty, honesty, and dependability which are common to a group. Thus, social capital cannot be acquired by individuals acting alone. It is created and transmitted through cultural mechanisms like tradition, religion, or historical habit, which create shared ethical values and a common purpose (Fukuyama, 1995, pp. 26-27).

The concept of social capital, as distinct from physical and human capital, facilitates market transactions (Coleman, 1988; Putnam, 1994). This occurs in three ways:

- the creation of a system of generalised reciprocity;
- the establishment of information channels, providing sorted and evaluated information and knowledge; and
- the simplification of market transactions by instituting norms and sanctions by which economic exchanges can occur, bypassing costly and legalistic institutional arrangements associated with market transactions.

Part of the motivation to create networks of firms is to foster shared values and trust-based relationships that go beyond purely market transactions. The shared aims of a network of firms, whether it is markets, training, or modernisation, generate new relationships that last well beyond the initial objective (Lipnack and Stamps, 1995). It is these social relationships and institutions in which the economic activity of a locality is embedded (Granovetter, 1992).

Before we claim to understand what it means for institutions to be *embedded*, we have to go beyond the simplistic, if important, dimension related to local sources of physical inputs (Dicken et al, 1995). Zukin and DiMaggio (1990) suggests that there are four kinds of embeddedness that affect economic action:

- *cognitive*, relating to disciplinary standards and ways of thinking;
- *cultural*, affecting broadly values and setting limits on market exchange;
- *structural*, referring to social structures and their effects - formal and informal - on economic activity;
- *political*, meaning the degree to which asymmetries of power affect economic activities and relationships.

Social capital seems to be the mechanism that makes places with 'institutional thickness' provide such a rich environment economic activity, noted by Amin and Thrift (1993, 1994). Putnam et al (1993) use the presence of social capital to explain the long-standing contrast between Northern and Southern Italy.

European models

The identification of industrial districts in the Third Italy prompted a groundswell of interest in interfirm relationships that go beyond material flows and include in fact several unmeasurable connections related to family ties, cultural closeness, and specialised knowledge. Piore and Sabel's (1984) book and examples of industrial districts and technology districts in Europe (e.g. Goodman et al, 1989; Maillat et al, 1995; Pyke et al, 1990; Storper, 1993) have prompted widespread attempts to create similar inter-firm collaboration in the USA (Rosenfeld, 1989-90; Bosworth and Rosenfeld, 1993; Lichtenstein, 1992; Lipnack and Stamps, 1993, pp. 135-159).

US networks: some background

All of the issues discussed thus far are relevant to the discussion of networks in the USA, despite the fact that the policies of network formation have been based on relatively few of them. In particular, the networks in Denmark and in Italy have been the principal models to which US networks have turned. Firm networks, structured to retain the best elements of both formal technology transfer programmes and informal interaction networks, are a recent, but growing, phenomenon throughout the country. Based explicitly on the industrial districts of Italy and Denmark (Rosenfeld, 1992), a variety of such networks have arisen in the USA since the late 1980s.

The term *flexible manufacturing networks* is perhaps the most common name for these (Bosworth and Rosenfeld, 1993; Pyke, 1994, pp. 113-115). Operating on a small geographic scale, and frequently involving smaller universities and community colleges (rather than major research university), firms in similar circumstances are encouraged to combine their capabilities in design, training, product development, and other activities. Quasi-public centers with a key coordinator or broker serve as the hub - and gatekeeper - for the network. The role of the broker - the common term in the USA for what Bennett and McCoshan (1993, pp. 210-211) call the 'core actor' - is considered critical (*Firm Connections*, 1994). Lipnack and Stamps (1993, p. 152) call brokers 'the spark plugs who guide the networks into existence'. Lichtenstein (1994) suggests that network facilitators, or practitioners of inter-firm collaboration, have several distinct roles. They must at various times act as sponsors, promoters, industry champions, brokers, scouts, coordinators, network administrators, and policymakers.

There is perhaps more diversity than commonality among the networks in place at this time; 27 such networks were included in a mid-1992 compilation, along with

201

a number of other 'network projects' being developed by state agencies, labour unions, and colleges and universities (Lichtenstein, 1992). Not all of these follow the guidelines suggested by the previous research reviewed earlier in this chapter; nor do they necessarily adhere to common sense generalisations, such as those proposed by Bosworth and Rosenberg (1993, pp. 4-6):

- networks are no panacea, especially when simply joining weak firms together;
- networks should exploit natural clusters of firms, where face-to-face interaction occurs naturally;
- network brokers, wherever they are housed, make a difference; and
- european models require translation.

The emerging US model, according to Bosworth and Rosenberg (1993, p. 29), is supported by local economic development agencies, building a multi-firm network organisation with a defined membership and internal structure. Although every activity does not involve every member, it is generally agreed that only members will participate, thus making this a *static* model. This contrasts with the *dynamic* model in Italy, where multiple network relationships evolve and dissolve almost organically.

Little previous research addresses the policy prescription of network creation in regions where networks have failed to emerge. Initially inspired by accounts by Piore and Sabel (1984), several networks have emerged in attempts to create the sorts of inter-firm cooperation found (especially) in Italy and Denmark. One protagonist was Hatch, whose ideas were promoted through the Corporation for Enterprise Development (Hatch, 1988). Others include Sommers (Northwest Policy Center, 1992) and Rosenfeld (1989-90) who, with Bosworth (Bosworth and Rosenfeld, 1993), were inspired by European examples. Through their energies, several conferences were held throughout the country in the late 1980s and 1990s, intended to publicise, and increase interest in flexible manufacturing networks. These conferences subsequently inspired several of the networks identified in a 1992 directory compiled under the auspices of the National Institute of Standards and Technology, or NIST (Lichtenstein, 1992). In the Pacific Northwest, Sommers and his colleagues have promoted experiments with networks (Rosenfeld, 1996; Sommers, 1994) and have widely disseminated information on manufacturing networks (Northwest Policy Center, 1992). Finally, among the states, Oregon has taken perhaps the largest step toward networking, through its Key Industries Program, designed in 1989 to increase international competitiveness in 27 clustered sectors (Oregon Economic Development Department, 1993; Regional Technology Strategies, 1993).

The diversity of networks is a result both of the diversity of local contexts and of the need for firms and other actors to learn about networks and about types and degrees of inter-firm collaboration. Bennett and McCoshan (1993) focus on public-private partnerships in which a development agency or other local actor begins the process of cooperation among all key agents. Bosworth's (1995) model is perhaps simpler, focusing on inter-firm collaboration with public policy 'intervention' required to promote the evolution from traditional industry associations to shared resource networks to co-production networks (Figure 9.1). Finally, Lichtenstein (1995) lists five essential collaborative services or products: (1) building awareness of inter-firm collaboration, (2) developing collaboration skills, (3) matchmaking, or linking one firm with a complementary firm, (4) implementing collaboratives with open-ended outcomes (soft networks), and (5) implementing collaboratives with specific outcomes (hard networks).

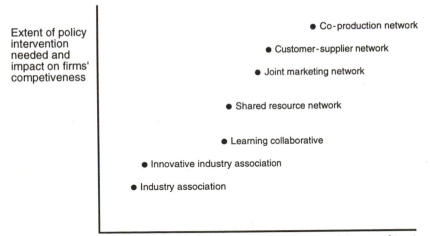

Figure 9.1 Types of interfirm collaboration

Source: Modified from Bosworth (1995).

US networks: empirical findings

Networks in the USA have attempted to respond to the need for modernisation of small and medium-sized firms (Shapira et al, 1995). Human capital as well as financial capital tend to be in short supply in small firms (O'Farrell and Hitchens, 1989). Small firms find that the demands of the market and the lack of resources (time as well as money) available to them keep them less than fully informed about potential markets elsewhere. Information about potential markets and channels for outputs is a great challenge to small enterprises, as is the struggle to maintain competitiveness as product and process innovation increase in importance. Networks make the entire process of information gathering easier and less costly for firms, an issue most significant to the newest and smallest firms.

Our research was based on the set of 27 flexible manufacturing networks identified as of late 1992 (Lichtenstein, 1992). Each of the networks, therefore, had been in existence for at least two years prior to contact from the research team. Of the 27 networks, ten are in metalworking industries (machining, metal fabrication), ten in woodworking (secondary wood products such as furniture), four in textiles, one in defense, one in plastics, and two in mixed sectors. Twenty-one of the networks list 'marketing' as an objective (Table 9.1).

Table 9. 1
Characteristics of flexible manufacturing networks in the USA

Industries		Network Objective	
Woodworking	10	Marketing	21
Metalworking	10	Training	10
Textiles and clothing	4	Production	8
Mixed industry	2	New product development	7
Plastics	1	Purchasing	6
Defense	1	Quality improvement	4

Source: Lichtenstein (1992), p. 3.

Networks continue to come into existence, promoted by the newsletter *Firm Connections* and, more recently, by a large-scale effort to create networks in 15 of the 50 states in the USA. This effort, dubbed USNet, is organised around a top-down model of broker training, prospective brokers being identified by state economic development or industrial extension agencies.

The initial research plan was to contact five networks in rural locations, five in urban locations, and five sets of firms not involved in network activity. This plan was largely followed, involving in-depth interviews with network brokers or coordinators of ten networks, and interviews with from two to seven member firms of each network. The firm interviews were conducted on the premises of the business, and took from 20 minutes to two hours, depending on the degree of discursive information (stories, history, anecdotes) provided by the owner-manager. The firm interview followed a questionnaire containing 31 items which probed various aspects of the firm's operation, as well as the firm's relationship with the network and with other firms in it. For the non-network firms, the questionnaire followed was shorter (about 20 items), omitting questions concerning network involvement. A total of 42 firms in networks were interviewed (24 in rural locations, 18 in urban networks). In addition, 19 non-network firm in three locations were interviewed to provide, a control group for inter-firm interaction.

The networks studied range from a small urban neighbourhood (the Metalworking Consortium, a part of the Jane Addams Resource Center in Chicago) to statewide efforts in Arkansas (the Metalworking Connection) and Louisiana (Louisiana Furnishings Industry Association) (Table 9.2). Two of the networks (Erie Bolt and the FlexCell Group) are private-sector operations, receiving no public funding of their broker's salary. In other respects, these two networks differ markedly. FlexCell members meet weekly to discuss individual strengths and common goals. Erie Bolt members rarely meet except as needed for a specific purpose. However, Erie Bolt has been in existence longer, and firms' mutual trust is based largely on the charismatic personality of Harry Brown, the president of EBC Industries, a maker of bolts and other metal fasteners. His belief in trust-based relationships has spread among firms in EBC's supplier network, bolstered by his larger, community links which have made funds for training available to other firms. In many respects, EBC is a 'strategic centre' of a 'web of partners' on a scale smaller than usually described (Lorenzoni and Baden-Fuller, p. 1995). More significantly perhaps, the Erie Bolt network exhibits characteristics of the dynamic networks found in Italy, with relationships fluid rather than fixed.

Several of the networks studied are based in community-development organisations which have added manufacturing networks to a wider existing array of services. The Metalworking Consortium , which serves a one mile by two mile neighbourhood on the north side of Chicago, is an example, as is ACEnet, an organisation which serves a large rural area of Ohio. For both, the network model seemed a useful addition to their activities that would help small firms to become more competitive. Administrators of both networks spend enormous amounts of time attracting funding for their organisation's activities, including network activities (Broun, 1994b, 1995).

Table 9.2
Networks included in the study

Network	Sector	Location	Number of firms interviewed
Accessible Housing Components (ACEnet)	wood and metal	Athens, Ohio rural	4
Erie Bolt	metal	Erie, Pennsylvania small urban	4
FlexCell Group	metal	Columbus, Indiana small urban	5
Louisiana Furnishings Industry Association*	wood	Ponchatoula, Louisiana rural	6
Metalworking Connection	metal	Arkadelphia, Arkansas rural	3
Metalworking Consortium	metal	Chicago urban	2
Philadelphia Guild	wood	Philadelphia urban	2
Technology Coast Manufacturing and Engineering Network (TeCMEN)	defense	Fort Walton Beach, Florida small urban	5
Wood Products Manufacturing Network	wood	Bemidji, Minnesota rural	4
WoodNet	wood	Port Angeles, Washington rural	7

* Originally called the Louisiana Furniture Industries Association

Some of the networks are fortunate to have community entrepreneurs as their coordinators. In several cases, surprisingly, the broker is a relative newcomer to the community, yet has been able to form links both among firms and, more importantly, with politicians and other sources of financial support. These links provide programmes, such as training, that would be difficult for small firms to arrange on their own.

The large-scale rural networks do not attempt to hold general membership meetings, although meetings of subsets of the group may take place. Instead, interaction tends to be initiated by the broker and sustained by mutual interest of some firms. This is especially true of the woodworking firms. To a large extent, the function of these dispersed networks is to provide services to firms, while at the same time initiating contacts among potential collaborators. Service provision is a critical element of small-firm competitiveness (Brusco, 1992). Vignettes of several of the networks will highlight the variety of represented.

The Appalachian Center for Economic Networks (ACEnet) is a nonprofit economic development organisation, based in Athens, Ohio, serving a rural 11-county region. ACEnet has functioned since the early 1980s, at first as an advocate for worker-owned enterprises. European models, especially the Basque Mondragon model, sparked an interest in flexible manufacturing and its cooperative emphasis. ACEnet's region is comprised almost entirely of micro-enterprises, and there are no large local firms on which small firms can rely. Consequently, a major objective of ACEnet has been the provision of several services it deems essential to develop an infrastructure of firms equipped to take advantage of the potential benefits of flexible production. Niche markets, including adjustable furniture and kitchens for wheelchair-bound residents, and food products for specific markets (organic, special diets, and ethnic foods) are the two primary areas of emphasis. Accessible Designs*Adjustable Systems (AD*AS) is a for-profit subsidiary under ACEnet's nonprofit umbrella. Of these, the Food Ventures (also known as Community Food Initiatives) is the more ambitious, with a much larger pool of potential clients being assisted toward self-employment. The services ACEnet offers its clients - largely comprised of single mothers with no business experience - include training in how to run a business.

With considerable financial support from the Ohio Department of Development, the Joyce Foundation, and the local business community, ACEnet has provided training to potential workers (enabling them to serve as apprentices in local firms), small amounts of seed capital, computer and bookkeeping skills, and telecommunications, and transitional support for welfare to self-employment. It has served as a model for other networks in other places in Ohio (Indergaard, 1996).

Firms with considerable experience also can benefit from networks, especially for dealing with a rapidly changing market and less predictable relationships with long-time customers. The FlexCell Group, in Columbus, Indiana, fits this description. This group developed out of the needs of a single firm client of the network facilitator. The original mission statement is one to which the other firms have also subscribed: *to become a full-service, vertically-integrated, single-source, top-tier network of suppliers servicing major manufacturers with product development and production capabilities.*

FlexCell has been a very dynamic network, with the number of 'core members' fluctuating and the need to build trust in the new members. There are presently five member firms, in addition to the consulting firm, which operates as a member. The network has been described previously as a strategic alliance (Lichtenstein, 1992, p. 20). These firms all focus on industrial customers, mainly, but not solely, for the automotive and industrial machinery industry. All members are 'job shops' which focus on custom work for much larger customers in the local vicinity (very much the situation described by Young et al, 1994). One firm describes its market as within 100 miles, but nearly all the firms have already begun to (or are just beginning to) sell to existing customers' facilities in Mexico.

The FlexCell Group formed in 1991, meeting weekly from that time and began to charge a regular fee of $US480 per month about a year later. This required degree of commitment prompted a firm to drop out, but other firms have also joined the network. A substantial commitment is demanded of the network's firms in addition to the CEO's (or a regular representative's) presence at weekly meetings. These meetings, as well as real efforts to get each firm to open up its operations, performance, and routines to the other members, are intended to develop trust among the members. The ultimate goal is to know enough about each other's operations and capabilities that any member can rely on the others to provide not only capacity but capability not present in any firm. The trust developed thus far has generated several instances of sharing facilities, jobs, and employees. Until mid1994, the network had received no public funding, making it different from the others described here. However, it was tapped as one of a small number of networks to receive US Department of Labor 'learning consortium' grants that will facilitate electronic communication and CAD/CAM links (Broun, 1994a).

The Wood Products Manufacturing Network, based in Bemidji, Minnesota, was one of five networks (WoodNet was another) funded for three years by the Northwest Area Foundation in 1991. Unlike WoodNet, the WPMN has attracted little additional funding and risks being absorbed at a much smaller scale of effort, into another state agency. The network has two active sub-networks of firms that have benefitted significantly from the WPMN. Associated Artists in Woods is a group of ten firms who are cooperating on a catalog of products to go to manufacturers' representatives for marketing. A second group of furniture makers has displayed products together at trade shows, participated in joint sales, and begun to work jointly to help the better firms in marketing to broaden their product lines.

The networks in the USA illustrate that no single model will apply. Courault and Romani (1992) and Bennett and McCoshan (1993) have come to a similar conclusion in Europe. The US networks conform to Mariti's (1990) characterisation that:

> cooperation between small firms tends to be confined to single activities and/or single products and/or single markets ... However, the content of the agreement tends to be 'complex' (more than one flow of products,

components or technology between the company up to the creation of common structures), though without reaching a great financial or managerial involvement. (p.43)

Some characteristics of firms in networks

The urban networks tend to have firms that are older and larger, as well as more 'high-tech' than those in the more rural networks. We know from membership lists that (rival) ACEnet is overwhelmingly comprised of small firms; 22 of 26 in a complete list have ten or fewer employees and a median size of four employees. A complete list for (urban) TeCMEN shows that its median size is 62.5 employees, with only one firm is in the 1-10 size group. Thus, the size distributions of interviewed firms are representative of their networks' populations. The Metalworking Connection in Arkansas is a typical network in terms of firm size; 55 of its 69 members (80 per cent) have from 1-10 employees, and only five firms are larger than 50 employees. Thus, many of the networks focus on services and advice geared toward new firms and micro-enterprises.

Two types of market orientation are important: export sales (indicating market competitiveness and non-dependence on the local market) and sales to firms in the network, which suggests local integration. Table 9.3 shows that the number of firms exporting ranges from 0 to 80 per cent (in FlexCell). The extent of sales going to export markets ranges from 0.5 per cent to 25 per cent. Six of the networks have considerable intra-network linkages (25 per cent-75 per cent of firms sell to other firms in their network). The range of intra-network sales ranges from 2 per cent to 75 per cent among the firms.

The firms are engaged in all forms of production, but most are into either custom or small-batch manufacturing. WoodNet firms, despite small firm sizes, are most involved in mass production. All FlexCell firms produce small batches; all ACEnet firms produce custom products. The majority of firms interviewed are strongly involved in new product development. Joint activities with other firms in the network are common. Joint production is particularly common (Table 9.4). In two of the networks, all firms interviewed have been engaged in some joint production with another network firm. Marketing and new product development are the most frequent activities, but training and purchasing, as well as technology development are also done.

Despite the varying local contexts - industrial sectors, economic development roles, size and number of members, funding sources - there are some generalisations about the services which networks provide to member firms. Networks provide at least seven categories of services: (1) meetings, (2) information clearinghouse functions, (3) business advice, (4) finance, (5) large-scale marketing efforts, (6) technical upgrading, and (7) training. These generally correspond to those discussed by Brusco (1992) and Schmitz and Musyck (1994) in European contexts. As Brusco puts it, services must be provided as a matter of policy because market failure

Table 9.3
Location of markets of firms in the ten networks

Network	% of firms exporting	Range of exports	% with sales to firms in network	Range of sales to firms in network
ACEnet	0.00	0.00	25	33%
Erie Bolt	25	0-0.5%	75	2-9%
FlexCell	80	5-25%	40	3-5%
LFIA	17	0-25%	0.00	0.00
Metalworking Connection	67	0.5-6%	0.00	0.00
Metalworking Consortium	0.00	0.00	0.00	0.00
Philadelphia Guild	50	0-12%	0.00	0.00
TeCMEN	40	2-17%	60	2-75%
Wood Products Manufacturing Network	25	0-1%	25	15%
WoodNet	71	0.5-25%	60	2-75%
All 42 firms	40	0-25%	26	2-75%

Table 9.4
Production collaboration by firms in the ten networks

Network	Per cent with joint production with other network firms	Range of joint production with other network firms
ACEnet	100table	1-100%
Erie Bolt	50	2-9%
FlexCell	60	1-12.5%
LFIA	67	1.5-20%
Metalworking Connection	67	0.5-3%
Metalworking Consortium	100	1-20%
Philadelphia Guild	0	0
TeCMEN	60	2.5-75%
Wood Products Manufacturing Network	75	1-30%
WoodNet	14	3%

typically prevents firms from providing them. Information about markets and technical standards are examples of things on which large firms are able to keep up, but which small firms need provided as a public good. Flexible networks serve just such a function - in rural and urban areas alike. Pyke (1994, pp. 83-84) stresses the difference between services for small firms and those for large firms. We found a similar distinction between networks that serve primarily new, small firms and those that serve established firms. The latter have less need for the full array of business advice, but often need technical upgrading or training services.

Networks: embedded institutions in regions

The creation of networks in places where there is no embedded structure for inter-firm collaboration is not a simple matter. Trust in the broker, in the concept of networks and, finally, in other firms must all be created from scratch, especially challenging when the network broker comes from outside the region. It also seems to be the case that some of the networks have difficulty in their evolution from one activity cycle, with one objective and set of resource needs, to another (Håkansson, 1992). A strong leader, such as that in the firm-based Erie Bolt network, may be able to ensure this better than can a diffuse, member-led organisation. Erie Bolt is one of the oldest networks in the USA, having been formed in 1987 (Richman, 1988).

The persistence of the TeCMEN network in Florida through three brokers and the spin-off of a competing network in just six years perhaps illustrates the embeddedness of networking among the 30 firms, as well as continuity of funding. The stability of the network in the absence of instability of its broker seems more to be a result of a deeply felt need among the firms for help in dealing with conversion from the shrinking military market to other, commercial markets. The similarity of the network's 30 firms in their dependence on the Pentagon generates a common purpose that transcends other pressures (Borfitz, 1994).

The role of the community entrepreneur is most significant. The most successful networks - judged by longevity and member enthusiasm - are those in which the broker or facilitator has taken on a role outside, and larger than, the network itself. Four of the ten networks have such a broker: ACEnet's is a long-time advocate of the poor in her region, a situation found also in the Metalworking Consortium's work in Chicago. WoodNet's broker invested enormous amounts of time to generate trust among his region's wood workers and to develop links to potential institutional support. Erie Bolt's broker is a key player in several public- and private-sector activities in his region, providing support for all small businesses in the locality, not just firms in his supplier network. This community spirit seems to be the hallmark of a community entrepreneur, who sees inter-firm networking as just one of several mechanisms for generating and maintaining local development. In some cases, such as Erie Bolt, links with other flexible manufacturing networks elsewhere are being formed. For example, EBC uses firms in the Heat-Treating Network in nearby Cleveland, Ohio, for specialised work.

The activities of the other networks are frequently single-focus, whether training (as in Arkansas), marketing (LFIA), or product development (Philadelphia Guild). This single focus seems to prevent embeddedness, because some firms will tend mainly to see the network primarily, or exclusively, as a provider of services, rather than as a resource to identify potential collaborator firms.

Conclusion

The descriptions and vignettes of flexible manufacturing networks illustrate the varied ways in which firms collaborate. Firms where networks are not present also collaborate, more informally, and they must do it without the watchful eye (and the fund-raising ability) of a network broker. For all firms, collaboration requires a substantial investment of time and money, and may be resisted by firm owners whose independence and 'fortress enterprise mentality' resist dependence on others (Curran and Blackburn, 1994). Collaboration also presumes some compatibility in culture, values, and objectives (Gertler, 1995). For firms not in network regions, firms often find their most fruitful knowledge and technical linkages at considerable distance (MacPherson, 1991, 1992; Malecki and Veldhoen, 1993). The degree of local embeddedness in such firms is minimal, of course, since their fortunes depend relatively little on the local economy.

It is very unclear whether networks of the type being formed in the USA will be able to constitute a milieu similar to those in Europe (Camagni, 1995; Maillat, 1990). As Bennett and McCoshan (1993) found in the UK, local cultures are difficult to change, and it is difficult to develop a culture of not only cooperation but of mutual trust and interdependence. Developing trust is an investment in longer-term change, and becomes a key part of the milieu and of successful cultures and economies (Fukuyama, 1995). In this sense, these US networks described in this paper are perhaps merely 'simple nodes' rather than 'complex nodes,' in Conti's (1993) typology.

It appears that networks which become embedded in their region tend to provide more for their member firms than do less well-integrated networks. Embeddedness is indicated by the involvement and financial (and other) support of local institutions and of influential residents, especially community entrepreneurs. In only a few regions where no network operates, firms are able nearly to substitute for a network by means of active informal networking on their own with firms from outside the locality. In most regions, some public-policy 'push' seems to be needed to foster networking among firms in the local area.

Our future work will examine more closely the differences, if any, between networks in urban and in rural settings, and to contrast these more systematically with each other and with non-networked firms, in order to assess the effectiveness of networks as an economic development strategy. The diversity found thus far, however, suggests that local embeddedness will be a major underlying factor that is likely to resist generalisations across networks and locales. Research of this type

213

brings to the fore the resourcefulness of small firms, and their ability (for the most part) to utilise their environments and their circumstances opportunistically. The presence of local 'community entrepreneurs' also adds to the local specificity, because these people and their ability to tap into local resources will take a large number of forms, depending on local history, culture, and 'institutional thickness' (Amin and Thrift, 1993, 1994). The efforts to foster networks is but one type of a large and diverse array of efforts to enhance or build local capacity (Bennett and McCoshan, 1993; Stöhr, 1990; Storper and Scott, 1992). The importance of understanding local specificity - and its generality - remains one of our biggest research challenges.

Acknowledgements

This research was supported by the US Department of Agriculture's Cooperative State Research Service National Research Inititiative Competitive Grants Program (NRICGP) Grant #93-37401-8989. Our sincere thanks to the many network coordinators and firm owners and managers who provided the information reported here.

References

Ahern, R. (1993a), 'The role of strategic alliances in the international organization of industry', *Environment and Planning A*, vol. 25, pp. 1229-1246.

Ahern, R. (1993b), 'Implications of strategic alliances for small R&D-intensive firms', *Environment and Planning A*, vol. 25, pp. 1511-1526.

Aldrich, H. and Zimmer, C. (1986), 'Entrepreneurship through social networks', in Sexton, D.L. and Smilor, R.W. (eds), *The Art and Science of Entrepreneurship*, Ballinger, Cambridge, MA, pp. 3-23.

Amin, A. and Thrift, N. (1993), 'Globalization, institutional thickness and local prospects', *Revue d'Economie Régionale et Urbaine*, no. 3, pp. 405-427.

Amin, A. and Thrift, N. (1994), 'Living in the global', in Amin, A. and Thrift, N. (eds) *Globalization, Institutions, and Regional Development in Europe*, Oxford University Press, Oxford, pp. 1-22.

Anderson, M. (1995), 'The role of collaborative integration in industrial organization: observations from the Canadian aerospace industry', *Economic Geography*, vol. 71, pp. 55-78

Axelsson, B. and Easton, G. (eds) (1992), *Industrial Networks: A New View of Reality*, Routledge, London.

Bennett R.J. and McCoshan, A. (1993), *Enterprise and Human Resource Development: Local Capacity Building,* Paul Chapman, London.

Borfitz, D. (1994), 'Cooperative capitalism', *Florida Trend*, vol. 36, no. 11, pp. 46-49.

Bosworth, B. (1995), 'Interfirm cooperation: The points of intervention', *Firm Connections*, vol. 3, no. 1, pp. 2, 5.

Bosworth, B. and Rosenfeld, S. (1993), *Significant Others: Exploring the Potential of Manufacturing Networks,* Regional Technology Strategies, Chapel Hill, NC.

Broun, D. (1994a), 'The FlexCell Group: salvaging the metal-cutting and pattern-making industry', *Firm Connections*, vol. 2, no. 5, pp. 3, 12.

Broun, D. (1994b), 'ACEnet in the hole: manufacturing networks with a twist', *Firm Connections*, vol. 2, no. 6, pp. 3, 12.

Broun, D. (1995), 'The Metalworking Consortium: wedding economic development to social service', *Firm Connections*, vol. 3, no. 1, pp. 3, 9.

Brusco, S. (1992), ' Small firms and the provision of real services', in Pyke, F. and Sengenberger, W. (eds), *Industrial Districts and Local Economic Generation*, International Institute for Labour Studies, Geneva, pp. 177-196.

Camagni, R. (1995), 'Global network and local milieu: towards a theory of economic space', in Conti, S., Malecki, E.J., and Oinas, P. (eds), *The Industrial Enterprise and Its Environment: Spatial Perspectives,* Avebury, Aldershot, pp. 195-214.

Casson, M. (1990), *Enterprise and Competitiveness: A Systems View of International Business*, Clarendon Press, Oxford.

Coleman, J. (1988), 'Social capital in the creation of human capital', *American Journal of Sociology*, vol. 94 (supplement), pp. S95-S120.

Conti, S. (1993), 'The network perspective in industrial geography: towards a model', *Geografiska Annaler*, vol. 75B, pp. 115-130.

Cooke, P. and Morgan, K. (1993), 'The network paradigm: new departures in corporate and regional development', *Environment and Planning D: Society and Space*, vol. 11, pp. 543-564.

Cromie, S., Birley, S. and Callaghan, I. (1993), 'Community brokers: their role in the formation and development of new ventures', *Entrepreneurship and Regional Development*, vol. 5, pp. 247-264.

Courault, B.A. and Romani, C. (1992), 'A reexamination of the Italian model of flexible production from a comparative point of view', in Storper, M. and Scott, A.J. (eds), *Pathways to Industrialization and Regional Development*, Routledge, London, pp. 205-215.

Curran, J. and Blackburn, R. (1994), *Small Firms and Local Economic Networks: The Death of the Local Economy?*, Paul Chapman, London.

Dicken, P. (1992), *Global Shift: The Internationalization of Economic Activity*, 2nd edition, Guilford Press, New York.

Eriksson, S. (1995), *Global Shift in the Aircraft Industry: A Study of Airframe Manufacturing with Special Reference to the Asian NIEs,* University of Gothenburg, School of Economic and Commerical Law, Gotheburg.

Dicken, P., Forsgren, M. and Malmberg, A. (1995), 'The local embeddedness of transnational corporation', in Amin, A. and Thrift, N. (eds), *Globalization, Institutions, and Regional Development in Europe*, Oxford University Press, Oxford, pp. 23-45.

Dubini, P. and Aldrich, H. (1991), ' Personal and extended networks are central to the entrepreneurial process', *Journal of Business Venturing*, vol. 6, pp. 305-313.

Easton, G. (1992), 'Industrial networks: a review', In Axelsson, B. and Easton, G. (eds), *Industrial Networks: A New View of Reality*, Routledge, London, pp. 3-27.

Easton, G. and Araujo, L. (1992),' Non-economic exchange in industrial networks', in Axelsson, B. and Easton, G. (eds), *Industrial Networks: A New View of Reality*, Routledge, London, pp. 62-84.

Falemo, B. (1989), 'The firm's external persons: entrepreneurs or network actors?', *Entrepreneurship and Regional Development*, vol. 1. pp. 167-177.

Firm Connections (1994), 'Building synergy: the hub as network catalyst', *Firm Connections*, vol. 2, no. 2, pp. 1, 4.

Fukuyama, F. (1995), *Trust: The Social Virtues and the Creation of Prosperity*, Free Press, New York.

Gelsing, L. (1992), 'Innovation and the development of industrial networks', in Lundvall, B.-A. (ed.), *National Systems of Innovation*, Pinter, London, pp. 116-128.

Gertler, M. (1993), 'Implementing advanced manufacturing technologies in mature industrial regions: towards a social model of technology production', *Regional Studies*, vol. 27, pp. 665-680.

Gertler, M.S. (1995), "Being there': Proximity, organization, and culture in the development and adoption of advanced manufacturing technologies', *Economic Geography*, vol. 71, pp. 1-26.

Goodman, E., Bamford, J. and Saynor, P. (eds) (1989), *Small Firms and Industrial Districts in Italy*, Routledge, London.

Granovetter, M. (1985), 'Economic action and social structure: the problem of embeddedness', *American Journal of Sociology*, vol 91, pp. 481-510.

Granovetter, M. (1992), 'Economic institutions as social constructions: a framework for analysis', *Acta Sociologica*, vol. 35, pp. 3-11.

Håkansson, H. (1989), *Corporate Technological Behaviour: Cooperation and Networks*, Routledge, London.

Håkansson, H. (1992), 'Evolution processes in industrial networks', in Axelsson, B. and Easton, G. (eds), *Industrial Networks: A New View of Reality*, Routledge, London, pp. 129-143.

Hansen, N. (1992), 'Competition, trust, and reciprocity in the development of innovative regional milieux', *Papers in Regional Science*, vol. 71, pp. 95-105.

Harrison, B. (1992), 'Industrial districts: old wine in new bottles?', *Regional Studies*, vol. 26, pp. 469-483.

Hatch, C.R. (1988), *Flexible Manufacturing Networks: Cooperation for Competitiveness in a Global Economy*, Corporation for Enterprise Development, Washington.

Hertz, S. (1992), 'Towards more integrated industrial systems', in Axelsson, B. and Easton, G. (eds), *Industrial Networks: A New View of Reality*, Routledge, London, pp. 105-128.

Hoare, A.G. (1985), 'Industrial linkage', in Pacione, M. (ed.), *Progress in Industrial Geography*, Croom Helm, London, pp. 40-81.

Illeris, S. and Jakobsen, L. (eds) (1990), *Networks and Regional Development*, NordREFO, Copenhagen.

Indergaard, M. (1996),'Making networks, remaking the city', *Economic Development Quarterly*, vol. 10, pp. 172-187.

Johannisson, B. (1990), 'The Nordic perspective: self-reliant local development in four Scandinavian countries', In Stöhr, W.B. (ed.), *Global Challenge and Local Response*, Mansell, London, pp. 57-89.

Johannisson, B. and Nilsson, A. (1989), 'Community entrepreneurs: networking for local development', *Entrepreneurship and Regional Development*, vol. 1, pp. 3-19.

Kanter, R.M. and Eccles, R.G. (1992), 'Making network research relevant to practice', in Nohria, N. and Eccles, R.G. (eds), *Networks and Organization*, Harvard Business School Press, Boston, pp. 521-527.

217

Leonard-Barton, D. (1995), *Wellsprings of Knowledge: Building and Sustaining the Sources of Innovation*, Harvard Business School Press, Boston.

Lewis, J.D. (1995), *The Connected Corporation: How Leading Companies Win Through Customer-Supplier Alliances*, Free Press, New York.

Lichtenstein, G.A. (1992) , *A Catalogue of U.S. Manufacturing Networks* (NIST GCR 92-616),MD, National Institute of Standards and Technology, Gaithersburg.

Lichtenstein, G.A. (1994), 'Training requirement for practitioners of inter-firm collaboration', Memo dated September 7, 1994, Valley Industrial Resource Center, Philadelphia, Delaware.

Lichtenstein, G.A. (1995), 'Collaborative services and products', unpublished paper.

Lipnack, J. and Stamps, J. (1993), *The TeamNet Factor*, Oliver Wright, Essex Junction, VT.

Lipnack, J. and Stamps, J. (1995), *The Age of the Network*, Oliver Wright, Essex Junction, VT.

Lorenz, E.H. (1992), 'Trust, community, and cooperation: toward a theory of industrial districts', in Storper, M. and Scott, A.J. (eds), *Pathways to Industrialization and Regional Development*, Routledge, London, pp. 195-204.

Lorenzoni, G. and Baden-Fuller, C. (1995), 'Creating a strategic center to manage a web of partners', *California Management Review*, vol. 37, no. 3, pp. 146-163.

Macdonald, S. and Williams, C. (1993), 'Beyond the boundary: an information perspective on the role of the gatekeeper in the organization', *Journal of Product Innovation Management*, vol. 10, pp. 417-427.

Macdonald, S. and Williams, C. (1994), 'The survival of the gatekeeper', *Research Policy*, vol. 23, pp. 123-132.

MacPherson, A. (1991),' Interfirm information linkages in an economically disadvantaged region: an empirical perspective from metropolitan Buffalo', *Environment and Planning A*, vol. 23, pp. 591-605.

MacPherson, A. (1992), 'Innovation, external technical linkages and small-firm commercial performance: an empirical analysis from Western New York', *Entrepreneurship and Regional Development*, vol. 4, pp. 165-183.

Maillat, D. (1990), 'SMEs, innovation and territorial development', in Cappellin, R. and Nijkamp, P. (eds), *The Spatial Context of Technological Development*, Avebury, Aldershor, pp. 331-351.

Maillat, D., Lecoq, B., Nemeti, F., and Pfister, M. (1995), 'Technology district and innovation: the case of the Swiss Jura Arc', *Regional Studies*, vol. 29, pp. 251-263.

Malecki, E.J. (1995), 'Flexibility and industrial districts', *Environment and Planning A*, vol. 27, pp. 11-14.

Malecki, E.J. and Veldhoen, M.E. (1993), 'Network activities, information and competitiveness in small firms', *Geografiska Annaler*, vol. 75B, pp. 131-147.

Mariti, P. (1990), 'Constructive cooperation between smaller firms for efficiency, quality and product changes', in O'Doherty, D. (ed.), *The Cooperation Phenomenon*, Graham and Trotman, London, pp. 31-50.

Nohria, N. and Eccles, R.G. (eds) (1992), *Networks and Organizations*, Harvard Business School Press, Boston.

Nonaka, I. and Takeuchi, H. (1995), *The Knowledge-Creating Company*, Oxford University Press, New York.

Northwest Policy Center (1992), *Entrepreneurial Strategies: Readings on Flexible Manufacturing Networks*, Northwest Policy Center, University of Washington, Seattle.

O'Farrell, P.N. and Hitchens, D.N. (1989), 'The competitiveness and performance of small manufacturing firms: an analysis of matched pairs in Scotland and England', *Environment and Planning A*, vol. 21, pp. 1241-1263.

Oregon Economic Development Department (1993), *Oregon Helping Oregon: A Progress Report on the Oregon Flexible Networks Initiative*, Oregon Economic Development Department, Industry Development Division, Salem.

Perrow, C. (1992), 'Small-firm networks', in Nohria, N. and Eccles, R.G. (eds), *Networks and Organizations*, Harvard Business School Press, Boston, pp. 445-470.

Phelps, N.A. (1992), 'External economies, agglomeration and flexible accumulation', *Transactions, Institute of British Geographers, New Series*, vol. 17, pp. 35-46.

Piore, M. and Sabel, C. (1984), *The Second Industrial Divide*, Basic Books, New York.

Putnam, R.D., Leonardi, R. and Nanetti, R.Y. (1993), *Making Democracy Work: Civic Traditions in Italy*, Princeton University Press, Princeton.

Putnam, R.D. (1994), 'The prosperous community: social capital and public life', *Firm Connections*, vol. 2, no. 2, pp. 5, 11-12.

Pyke, F. (1992), *Industrial Development through Small-Firm Cooperation*, International Labour Office, Geneva.

Pyke, F. (1994), *Small Firms, Technical Services and Inter-Firm Cooperation*, International Institute for Labour Studies, Geneva.

Pyke, F., Becattini, G. and Sengenberger, W. (eds) (1990), *Industrial Districts and Inter-Firm Co-operation in Italy*, International Institute for Labour Studies, Geneva.

Regional Technology Strategies (1993), *Oregon's Key Industries Program: Assessing the Impact of Manufacturing Networks*, Regional Technology Strategies, Chapel Hill, NC.

Richman, T. (1988), 'Make love, not war: how competitors are joining forces to create opportunities they wouldn't get on their own', *Inc.*, vol. 10, no. 8, pp. 56-60.

Robertson, P.L. and Langlois, R.N. (1995), 'Innovation, networks, and vertical integration', *Research Policy*, vol. 24, pp. 543-562.

Rosenfeld, S. (1989-90), 'Regional development European style', *Issues in Science and Technology*, vol. 6, no. 2, pp. 63-70.

Rosenfeld, S. (1992), *Competitive Manufacturing: New Strategies for Rural Development*, Center for Urban Policy Research Press, Piscataway, NJ.

Rosenfeld, S. (1996), 'Does cooperation enhance competitiveness? Assessing the impacts of inter-firm collaboration', *Research Policy*, vol. 25, pp. 247-263.

Sabel, C.F. (1992), 'Studied trust: building new forms of co-operation in a volatile economy', in Pyke, F. and Sengenberger, W. (eds), *Industrial Districts and Local Economic Regeneration*, International Institute for Labour Studies, Geneva, pp. 215-250.

Sako, M. (1992), *Prices, Quality, and Trust: Inter-Firm Relations in Britain and Japan*, Cambridge University Press, Cambridge.

Saxenian, A. (1994), *Regional Advantage: Culture and Competition in Silicon Valley and Route 128*, Harvard University Press, Cambridge, MA.

Schmitz, H. and Musyck, B. (1994), 'Industrial districts in Europe: policy lessons for developing countries?', *World Development*, vol. 22, pp. 889-910.

Shapira, P., Roessner, J.D. and Barke, R. (1995), 'New public infrastructure for small firm industrial modernization in the USA', *Entrepreneurship and Regional Development*, vol. 7, pp. 63-84.

Sommers, P. (1994), 'Manufacturing networks show progress', *The Changing Northwest*, vol. 6, no. 1, pp. 1, 6.

Stöhr, W. (1990), 'Synthesis', in Stöhr, W.B. (ed.), *Global Challenge and Local Response*, Mansell, London, pp. 1-19.

Storper, M. (1993), 'Regional 'worlds' of production: learning and innovation in the technology districts of France, Italy and the USA', *Regional Studies*, vol. 27, pp. 433-455.

Storper, M. and Harrison, B. (1991), 'Flexibility, hierarchy and regional development: the changing structure of industrial production systems and their forms of governance in the 1990s', *Research Policy*, vol. 20, pp. 407-422.

Storper, M. and Scott, A.J. (eds) (1992), *Pathways to Industrialization and Regional Development*, Routledge, London.

Sweeney, G.P. (1991), 'Technical culture and the local dimension of entrepreneurial vitality', *Entrepreneurship and Regional Development*, vol. 3, pp. 363-378.

Thorngren, B. (1970), 'How do contact networks affect regional development?', *Environment and Planning*, vol. 2, pp. 409-427.

Vaessen, P. and Wever, E. (1993), 'Spatial responsiveness of small firms', *Tijdschrift voor Economische en Sociale Geografie*, vol. 84, pp. 119-131.

Vatne, E. (1995), 'Local resource mobilization and internationalization strategies in small and medium sized enterprises', *Environment and Planning A*, vol. 27, pp. 63-80.

Yeung, H. (1994), 'Critical reviews of geographical perspectives on business organizations and the organization of production: towards a network approach', *Progress in Human Geography*, vol. 18, pp. 460-490.

Young, R.C., Francis, J.D. and Young, C.H. (1994), 'Flexibility in small manufacturing firms and regional industrial formations', *Regional Studies*, vol. 28, pp. 27-38.

Zukin, S. and DiMaggio, P. (1990), 'Introduction', in Zukin, S. and DiMaggio, P. (eds), *Structures of Capital: The Social Organization of the Economy*, Cambridge University Press, Cambridge, pp. 1-36.

Part III: Labour, locality and restructuring

Central to the problem of global-local interdependence is how localities and industrial districts can survive and succeed in a globalised economy that appears to undermine them, denies them autonomy and generally recasts localities and places as disempowered victims. The three chapters of this part of the volume address this issue of the locality as the victim of globalisation in the context of labour, locality, corporate restructuring and discourses of control. Fagan views this issue from the perspective of peoples' local labour market experiences in the western suburbs of Sydney and points to the ideology of globalisation as a mechanism of change and control. Jonas provides further insights into labour control under conditions of globalisation in the context of case studies from the USA. Finally, O'Neill develops a discourse of control that surrounds the restructuring of BHP's 'mother plant' in Newcastle, New South Wales, comprising corporate rhetoric, restructuring and the realisation of residual value.

10   Globalisation and the suburbs: local labour markets in Western Sydney

Robert Fagan

Introduction

In the 1990s, Sydney has been characterised increasingly by local politicians and media commentators as a global city. Indeed, the notion has received some academic support (Daly and Stimson, 1992), with Searle (1996, p.11) arguing that Sydney has become a global city '... to a significant degree' and Sassen (1994, p. 4) confirming that the city is Australia's only candidate for the label. The concept of globalisation has become a crucial dimension in identifying global cities. While recognising the importance of the particularities of place, Sassen (1994) looks for a geography of 'strategic places' on a global scale, characterising them as command centres for organisation of a new international economy underpinned by rapid flows of information - a new 'informational capitalism' (Luke, 1994, p. 619) underpinned by information technology. This economy is organised increasingly around a global financial system, has featured rapid increases in the volume of both world trade and international flows of money, and has experienced major changes in the organisational forms and behaviour of transnational corporations (TNCs).

According to Fainstein, Gordon and Harloe (1992) and Sassen (1994, p. 99-118), new patterns of inequality have emerged in global cities alongside, or partly as a result of, their global functions. In this literature, rediscovery of inner city living by a new middle class of urban workers, mostly in the information industries, has focused attention on gentrification since 1980. In addition, redevelopment by global finance capital (the imagery of the London docklands or Sydney's Darling Harbour) is seen to stand alongside remnants of inner city working class areas, often with high concentrations of particular ethnic groups, and to contrast with uneven impacts of economic restructuring contested by emerging urban social movements. This has created an alternative urban politics focused around: first, impacts of both restructuring and globalisation; second, inner city urban environmental impacts (freeways, heliports - or, in Sydney's case, airports - loss of public open spaces); and third, contested representations of both gender and sexuality.

Yet the suburbs have been neglected almost entirely in this literature. Suburbanisation of population and economic activity has continued unabated over this same period in all of the global cities. Inner areas have long been net losers of population and economic activity, not just of manufacturing but now a wide range of service and office-based activities. For increasing proportions of these cities' residents, outer suburbs are the sites of both their residential and work-place experiences. In Britain and the United States, rapid suburbanisation of economic activity and middle class (white) populations has often been linked directly to the marginalisation of people in the inner city areas. No doubt inner New York, with its 'ghetto poor' (Fainstein and Harloe, 1992, p. 11) bound up in complex ways with racial prejudice, and inner London, with its 'tenement city' based on public housing (Harloe and Fainstein, 1992, p. 262), are major sites of social and political marginalisation in those so-called global cities. Yet none of the detailed analyses in Fainstein, Gordon and Harloe (1992) discuss the possibility of suburban marginalisation.

There is no such neglect of suburbs in the fields of feminist and cultural geography. According to Healy (1994, p. xiii) the concept of the suburb is similar to that of 'culture' in general, '...both can denote specific social practices or a way of life: [they] can evoke the mundane or the quintessence of the human spirit'. This recent view stands in contrast to long-established views in urban political economy which interpret suburbs mechanistically, largely as necessary consumption spaces for expanded capital accumulation, as emerging residential norms in the reproduction of the labour force and, largely through the operation of housing markets, as part of the state's legitimation goal of ensuring a stake in the system for the urban working class. Feminist analyses have demonstrated the crucial link between the highly gendered notions of public and private spheres and the city-suburb dichotomy. Women and suburbs share the 'domestic' or private in contrast to the downtown world of men and the power structures of the public sphere (capital and state). Proliferation of single-family homes in suburbs, especially after 1950, reinforced the 'domestic' roles of women, spatially separating them from the city's concentrations of workplaces (Hanson and Pratt, 1995, p. 94)

More recent analyses (see, for example, Ferber, et al 1994; Gibson and Watson, 1994b) reveal sharply differentiated views of the suburbs in which 'top-down' processes of restructuring, involving either capital or state, are contested as well as accommodated, and in which '...women's views can be seen as both participating in and challenging established representations which...map the suburbs onto the feminine, the materialistic and the domestic'. (Gibson and Watson, 1994a, p. 14). In any case, since the onset of economic restructuring in the 1970s, suburban women have entered the paid workforce in record numbers. While Hanson and Pratt (1995) show greater 'localism' among women in their contemporary experiences of waged-work, increasing numbers of men travel within the suburbs to work in suburbanised factories while women commute to the CBDs of global cities in greater numbers (especially young office workers). Further, the home is the site of waged-work for increasing numbers of suburban women through their participation in outworking (for example in the clothing industry), while both women and men work in the information economy from home-based 'work-stations'. Class, gender, ethnicity and age all interact with specifics of place to construct differential access to employment opportunities in the suburbs.

Understanding suburban labour markets requires incorporating the impacts of globalisation while avoiding the deafening silences about the suburbs in the analysis of global cities. A recent voluminous survey of globalisation by Barnet and Cavanagh (1995), for example, makes only a single reference to the suburbs and then only in their context as alternative sites for corporations seeking lower cost locations for factories or administrative offices, either tapping into new pools of, mostly female, labour in what they describe revealingly as 'bedroom communities' (1995, p. 277), or attracted to other localised site advantages such as cheap land or lower taxes. Yet this chapter also seeks to extend the much richer examination of suburbanisation and localism offered by Hanson and Pratt (1995) who make only passing reference to globalisation, pointing out that '... attention [given] to the globalization of culture and economy should not allow us to lose sight of the rootedness of local lives' (1995, p. 22).

One key to such avoidance and extension is to recognise globalisation not only as a complex set of economic and social reconstructions but also as a powerful political and ideological force at national and local scales. The theoretical and empirical turn towards the global since 1980 has coincided with, and partly created, a proliferation of *global metaphors* - global city joins notions of global factory, global market and global workplace. All of these are abstractions which can bury the continued importance of national, regional and local scales of analysis. They often exaggerate both the degree of economic integration and the determination of the system by forces coming from the top down. This has reinforced an over-arching 'steam-roller' model of globalisation; changes take place 'out there', for example in the Pacific Rim, and must be accommodated 'in here' - for example, Australia's cities and exporting regions. Hence, globalisation is both a process and a discourse (see also Koc, 1994). Marcuse (1995) distinguishes between 'real' globalisation, which traces from place to place the different combinations and uneven impacts of technological change, global economic integration and decline of state power, and 'glossy' globalisation with which powerful stakeholders attempt to conceal this unevenness and the conflicts which underlie it. While this distinction is problematic theoretically, it points to the powerful ideological dimension and questions the role of nation-states both in the globalisation process and in sustaining the globalisation discourse.

This chapter seeks some understanding of people's contemporary suburban employment experiences through a case-study of outer suburban labour markets in Sydney. Western Sydney (Figure 10.1) has experienced considerable economic and demographic growth since the mid 1980s while maintaining unemployment rates well above metropolitan averages over the same period. This chapter shows that for outer

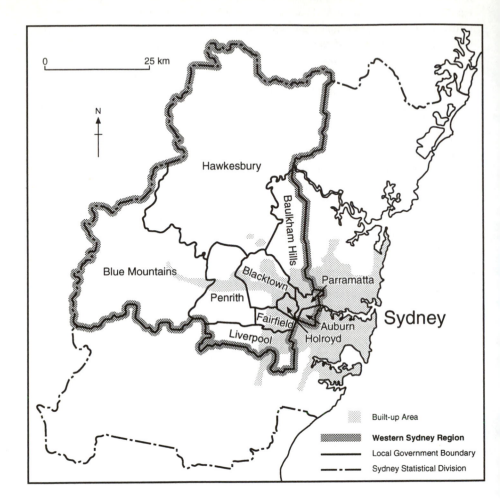

Built-up Area

Western Sydney Region

Local Government Boundary

Sydney Statistical Division

Figure 10.1 Western Sydney and the Sydney Statistical Division

Source: Fagan, 1986

suburban areas of cities like Sydney, the ideology of globalisation has become as important as measurable impacts of global change because of its key role in sustaining major shifts in economic and social policies of the state. The chapter concludes by considering local employment access, a stronger basis for understanding the articulation of global and local in suburban labour markets.

Australian employment policy and globalisation

The debate over globalisation in academia has strong parallels in the political sphere. Globalisation has become one keystone of a dominant discourse within the powerful policy-making cultures ascendant in most industrialised countries since the early 1980s. Global imperatives have become part of the rationale for complex processes which underlie major shifts away from regulation strategies implemented by the state during the so-called Fordist long boom - although not necessarily towards deregulation as such. Yet in their industry and employment policies, nation states have been driven to a greater extent by a particular ideology of globalisation than by measurable impacts of global change on their local industries and industrial places.

By the early 1980s, the economic, social and political pillars which supported post-1950 industry and employment policy in Australia had crumbled. Yet there, as elsewhere among older industrialised countries, the need to respond to globalisation gained a narrow political expression as the requirement that domestic economic activity meet international standards of profitability and adopt 'international best practice'. Indeed, during the period of federal Australian Labor Government (1983-1996) there was very limited debate in Australia about industry policy frameworks that did not lie within the paradigm of international competitiveness. The conservative coalition elected to government in March 1996 also pitched its economic policies in this paradigm but its rhetoric implied even less state intervention in the globalisation processes.

For more than a decade, then, internationalisation has been posed as a constraint on domestic activity at the macro-economic level, epitomised by a focus in both politics and the media on the balance of payments as a critical barrier to Australia's economic progress. Economic policy has been aimed at increasing exports at a rate faster than the growth of imports, almost in spite of the actual state of Australia's external accounts, dominated as they have been since the early 1980s by a net deficit on trade in services and by a very large net income payable overseas (mostly interest payments on debt raised by the private sector from global banks).

This public policy conception reflects the steam-roller ideology outlined earlier. The global economy must be defended against, or (hopefully) brought into fruitful partnership, via the extension of exports and by maximising domestic import-replacement, but at prices reflecting world markets. Ironically the recent adoption by senior Australian government advisers of the notion of encouraging manufacturing with demonstrable *competitive advantage* in the new markets of Pacific Asia, such as food processing, followed the work of Michael Porter (1990). Yet they appear to have ignored Porter's central argument that actual market competition is about corporate planning and strategy and, internationally, about competitive state policies and the creation of business environments conducive to investment through national industry strategies. It does not arise from a meta-theoretical process of market determination.

Similarly, the steam-roller metaphor underpinned the Labor Government's re-discovery of the regions in their employment policy adopted in 1994. During the early 1990s, the reappearance of double-digit national unemployment figures led to public commitments by the Government to introduce more substantial employment policy. Further, future electoral vulnerability was recognised in the highly uneven geography of unemployment, especially long-term joblessness. One response was a series of inquiries into the regions which included: a high profile inquiry chaired by a national leader of the trade union movement (Taskforce on Regional Development, 1993); an inquiry within the federal bureaucracy into impediments to regional industry adjustment; and a report on business investment in 'regional Australia' by international consultants McKinsey and Company (1994). *Working Nation*, released after these inquiries, included specific regional policies but was largely an endorsement of the government's internationalisation strategy, seeking to encourage globally competitive industries to maintain national economic growth-rates which would deliver jobs.

A detailed analysis of Australia's recent employment policy is beyond the scope of this chapter. (For more details, see *Australian Geographer*, 1994.) Further, the newly-elected conservative coalition pledged to abandon *Working Nation* and make major changes to Australia's industrial relations system, with polices to help small business growth, rather than retain directed labour market programmes. Soon after its election, the Liberal-National Government announced the formal abandonment of federal regional policy. It is significant, however, that both *Working Nation* and the polices of the new government effectively endorse findings of the McKinsey inquiry which stressed the importance to job creation of local 'leadership' and entrepreneurialism. McKinsey and Company had also encouraged the Federal Government to assist local firms in adopting international 'best practice' in their technologies, management and industrial relations. (Which countries constitute the exemplar for these industrial relations was not specified.) *Working Nation's* regional strategy encouraged regions to lift themselves out of their own employment difficulties - a 'bottom up' approach involving places getting to know their own strengths and weaknesses and mobilising their local private sector.

Yet at least three problems were overlooked in this approach. First, the McKinsey Report paid no attention to either the Australian or overseas track-record of local employment initiatives which were widely debated throughout industrialised countries during the 1980s (Eisenschitz and Gough, 1993). While genuinely local initiatives were sometimes effective ways of mobilising community action, they faced enormous political and financial obstacles. Whether they were locally progressive or regressive depended on contingent circumstances especially national and local politics (Beynon and Hudson, 1991). Local initiatives developed from the 'top-down' were often political smokescreens for effective withdrawal by the state from regional policy, delivery vehicles for changes in social welfare programmes, or starting gates for a competition between places for new private sector investment.

Second, despite welcome recognition that people experience labour market changes locally (Australia, 1994b, p. 17), the concept of locality did not inform the white paper's strategic thinking. There was no recognition of ways in which local forces shape either the availability of waged-work or people's *access* to jobs. Regions were little more than convenient (or historical) ways of dividing up territory. Third, there was little or no understanding of relationships *between* scales; international best practice, 'going global' (Australia, 1994a, p. 56), and building local business (read internationally competitive firms) were juxtaposed in the strategy with no concern for global-local interactions. Finally, despite references to difficulties faced by local areas within Australia's metropolitan cities, including Western Sydney, *Working Nation's* regional strategy was centred clearly on non-metropolitan or peripheral Australia which had been the spatial reference for the McKinsey study.

Globalisation and western Sydney's labour markets[1]

Sydney's outer western suburbs present a series of contradictions. On the one hand, more than 70 per cent of net population growth in the Sydney metropolitan area between 1981 and 1991 was accounted for by the western suburbs. Economic activity continued to suburbanise during the 1980s and Western Sydney increased its relative share of Australia's declining employment in manufacturing. This suburbanisation is not adequately represented by the Barnet and Cavanagh model of capital seeking out lower cost locations against the background of globalisation and rationalisation in the inner city. Much of the activity is 'indigenous' to the outer suburbs growing to serve local markets or new distributional and service activities aimed at regional or state-wide markets. As well as these proliferating small businesses, some of Australia's most powerful corporations, including the branch plants of TNCs, maintain their major processing plants in the western suburbs, established during the 1960s and 1970s to take advantage of Sydney's centrality to sell products in both national and export markets. At selected outer suburban business centres, office employment grew rapidly and most job growth in the region mirrored the national trend being concentrated in commercial and personal services.

On the other hand, severe job-shedding took place in key manufacturing sectors even on the industrial estates which had grown rapidly during the 1980s. Unemployment rates, which reflected falling national trends from 1984 to 1990, accelerated across most of the outer suburbs after 1990, some local government areas (LGAs) recording among Australia's highest rates by 1994 (Figure 10.2). In the Fairfield-Liverpool area of the southwestern suburbs, for example, an overall unemployment rate of 14.6 per cent in January 1992 (compared with 9.6 per cent nationally) stood at 39.0 per cent for people between the ages of 15 and 19 (21.0 per

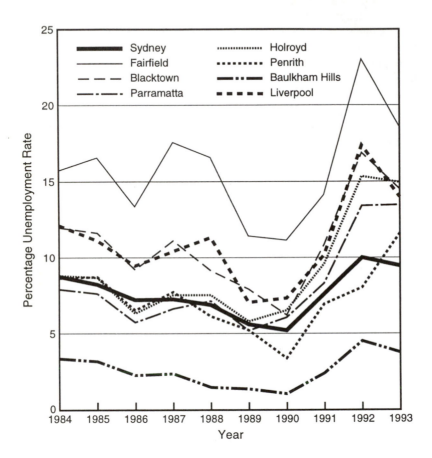

Figure 10.2   Western Sydney : Unemployment rates - 1984-1993

Source: Australia, Department of Employment, Education and Training, *Small Area Labour Markets - Australia,* AGPS, Canberra, 1994

cent nationally). Residents in many outer suburbs fell further behind metropolitan averages in their local access to a wide range of social services, educational facilities, public transport and other urban amenities. For other outer suburban residents, however, levels of socio-economic well-being looked little different from those in moderately well-off suburbs to the northwest and south of the city. This jigsaw puzzle of population growth, rapid industrial growth and decline makes outer Western Sydney a complex set of places for people to live and work.

Local labour markets in Western Sydney are extremely open, part of Sydney's wider metropolitan labour market demarcated conventionally by daily commuting patterns. Yet Hanson and Pratt (1995) show that these definitions are inadequate, bringing together widely differing labour market experiences and often dominated by the journey-to-work behaviour of male workers. People always experience a variety of overlapping local labour markets reflecting their differential access to employment opportunities which is constructed both socially and economically and shaped as much by home as workplace. In a detailed study of Worcester (Massachusetts), they showed that 'localism' was not an intrinsic characteristic of women in labour markets but was related to complex interactions between patriarchal social structures, class and geography (Hanson and Pratt, 1995, p. 226), no one of which is deterministic.

Massey (1993) has argued for localities as communities of interest - a complex of overlapping places depending on the standpoints of residents and institutions. Such localities exist in the region delimited by Western Sydney Regional Organisation of Councils (Figure 10.1) representing nine Local Councils (LGAs) which have for nearly two decades shared a community of interest in the high unemployment rates despite major economic and social differences between them. These communities have also been constructed *culturally* by shared regional consciousness among their residents, but also by people living elsewhere in the metropolis, often on the basis of images and stereotypes. Outer suburban residents, especially the young, continue to suffer from stereotyping of both people and places out west. The complex patchwork of economic and social circumstances that shape the outer suburbs is often buried beneath the pejorative description of its residents as 'westies'. In many ways, the western suburbs have been constructed culturally as Sydney's 'other' (Hodge, 1993). Hence, people's local labour market experiences are fluid and are partly constructed by social, economic and cultural differentiation within the suburbs.

Demographic change

Demographic change continues to underlie suburbanisation of urban society and economy. Western Sydney (as defined in Figure 10.1) housed 1.2 million people in 1991. Population growth between 1986 and 1991 varied from 18 per cent in rural-urban fringe areas like Hawkesbury, to 14 per cent in Fairfield and more than 10 per cent in Baulkham Hills, Blacktown and Penrith. On the inner western side in Holroyd

233

and Parramatta, however, growth was just over 1 per cent, compared with the Sydney metropolitan change of 5 per cent. Blacktown alone housed 212,000 people in 1991, Fairfield 175,000 and Penrith 150,000. In common with the metropolis as a whole, Western Sydney is now a net loser of population through internal migration. More people left the outer suburbs between 1981 and 1991 than moved in from other parts of Sydney or elsewhere in Australia, reflecting both Sydney's net loss of population to other places since the mid 1970s and households moving closer to the CBD for employment reasons or by 'trading up' in the housing market. Within the West, however, there has been a cascading effect with outer LGAs still receiving net in-migration from inner western LGAs. Natural increase has become the principal component of outer suburban population growth since rates of new family formation remain high and there is a high proportion of children in the fast-growing LGAs.

Yet migration from overseas continues to contribute directly to these high rates of population growth in Western Sydney. Since 1986, Sydney has become Australia's principal destination for overseas migration, especially from the Asia-Pacific Region. The outer western suburbs now contain some of Sydney's largest concentrations of immigrants from non-English-speaking backgrounds (NESB), swelled both by new arrivals and NESB people relocating from inner city areas. Between 1986 and 1991, new migrants from NES countries, especially from the Middle East and Pacific Asia, were concentrated in Fairfield, Blacktown, Liverpool and Parramatta while below metropolitan average concentrations were found in Baulkham Hills, Penrith, Holroyd, Blue Mountains and Hawkesbury. In 1991, whereas 26 per cent of the Australian work-force comprised NESB people, this percentage was 35 in Liverpool and 50 in Fairfield. Looked at another way: the receipt of overseas migrants would have made a much greater contribution to outer suburban population growth were it not for the relatively high out-migration of people (much less likely to be NESB) by the 1990s. Between 1986 and 1991, Sydney received 236,000 arrivals from overseas but the total net addition to the population from all migration was 90,400. Hence 146,000 people left Sydney over this period (Burnley and Murphy, 1994).

Continued suburbanisation of Sydney's population overall, therefore, reflects: natural increase and overseas migration tempered by net population loss through out-migration. This demography builds on populations which grew on the fringes initially because of the operation of metropolitan residential land and housing markets; the location of low-cost public housing in outer suburbs; the continued migration of well-off, professional households into formerly working class housing areas around the inner city, displacing such groups to lower cost housing areas in the suburbs; the small in-roads made into urban sprawl by the State Government's urban consolidation policies; and the clustering of NESB groups in particular outer suburban localities. The youthful nature of the labour force will clearly underpin a high growth-rate in the demand for employment well into the next century. Yet this suburban population growth, in the past, has stimulated new employment in the suburbs and does not, by itself, explain the puzzle of the West's high jobless rates.

Restructuring the suburban economy: global and local forces

Given the high rates of growth of the labour force since the early 1970s, commuting has always played a major part in getting outer suburban residents into waged-work. While this remains crucial for some people in some parts of the West and in some sectors (especially for office jobs), there has been steady suburbanisation of economic activity within Sydney since the mid 1960s. The growing population has provided the principal market for a wide range of manufactured goods and services with the 'centre of gravity' of Sydney's manufacturing shifting steadily westwards. Yet globalisation has also had a major impact on this outer suburban manufacturing economy. Western Sydney has not been immune from the national deindustrialisation of the labour force, a phenomenon which also began to suburbanise after 1980. After widespread industrial growth during the 1970s, except on the innermost industrial estates near the head of the Parramatta River, there was a net loss of manufacturing jobs in virtually all outer suburban LGAs between 1980 and 1984. Even Penrith, with some of Sydney's most recently-developed industrial estates, lost more than 1,000 manufacturing jobs in a total of nearly 51,000 jobs shed from Western Sydney's factories.

There were five principal reasons, only two of them 'home-grown'.

(1)   Despite continued suburbanisation of the market, the growth rate of new businesses to serve it was curtailed by recession, especially in the building products industries. Existing firms across all manufacturing sectors experienced technological change, and rationalisation during the 1980s, and there was considerable intensification of the labour process with the growth of sub-contracting and some casualisation of the factory workforce.

(2)   Relocation of manufacturing from other parts of the metropolis throughout the 1970s, often onto 'greenfield' sites, had effectively run out of steam by 1981.

(3)   The prevailing national strategy of internationalisation also had a direct impact on Western Sydney during the 1980s. Deregulation of the macro-economy was aimed at further integrating Australia with the global economy. While some of Western Sydney's manufacturers were successful exporters, especially to the Asia-Pacific region, industrial production in the outer suburbs was oriented primarily towards national and local markets, especially in food processing, metal fabrication and heavy engineering. Because of the sectoral composition of industry, effective levels of tariff protection for Western Sydney's manufacturing were already low. Yet, while the direct impacts of accelerated tariff reduction after 1984 were not substantial (they had a much greater impact on manufacturing in Melbourne), Sydney's outer suburban industrial production has proven

235

quite vulnerable to changing trade patterns throughout the Asia-Pacific region and the growth of NICs. Increased import competition affected TCF production while many larger manufacturers of industrial, transport and electrical equipment effectively disappeared from Western Sydney in the 1980s.

(4)   Changing corporate strategies affected many of Western Sydney's larger plants, especially those owned by TNCs. Some plants became warehouses for products imported from overseas parents, others lost their share of export markets in Asia to a combination of lower-cost production in NICs and various increases in (non-tariff) protection of these markets, while still others were closed as part of national rationalisation and centralisation to other capital cities such as Melbourne.

(5)   Western Sydney's largest remaining manufacturing sector is food processing. Food and beverage manufacturing has been affected strongly by global restructuring and decisions made by the large Australian-owned and transnational firms which dominate the local industry. Most of large producers carried massive burdens of debt into the 1990s. Firms had borrowed from the global financial system to finance domestic or global acquisitions and major offshore expansions. Major restructuring and job-shedding followed at the local level, some of it aimed at securing Australian market share against local rivals or increased imports, but all of it financially-driven as parent companies struggled to restructure their debt. For many food plants in Western Sydney, these financial uncertainties at national or global levels caused new local investment strategies to become stalled by the early 1990s.

Despite these changes, Western Sydney has continued to develop as one of Australia's most important manufacturing regions since 1980. Between 1985 and 1989 there was a net addition of 7,675 manufacturing jobs distributed across most sectors but concentrated in plants manufacturing paper products, plastics, wood products, processed food and clothing (Table 10.1). Yet the vast proportion of these new manufacturing jobs created after 1985 were in just three LGAs, Fairfield, Liverpool and Blacktown (Table 10. 2).

The local impacts of these changes emanating from the national and global scales since 1980 have been determined partly by local economic and labour market circumstances. These figures record *net* changes - the number of new jobs each year minus the jobs which disappear. While the balance has been positive in Western Sydney since the mid 1980s, jobs have been shed typically from large factories often involving workers with higher skills. Jobs have grown most often in small firms which continue to proliferate in the outer suburbs to serve rapidly-growing local markets and sub-contracting arrangements. These firms have commonly offered work

Table 10.1

Employment changes in manufacturing - Western Sydney, 1984-85 to 1988-89

SECTOR	1984-85	1988-89	Change	Percent
Food, beverages	12,663	14,765	+2,102	+16.6
Textiles	2,271	1,793	-478	-21.0
Clothing	3,863	4,346	+483	+12.5
Wood products	6,795	7,557	+762	+11.2
Paper products, printing	4,804	6,360	+1,556	+32.4
Chemicals, oil	8,115	8,725	+610	+7.5
Non-metallic minerals	5,083	5,418	+335	+6.6
Basic metals	4,906	5,038	+132	+2.7
Metal fabrication	10,824	11,817	+993	+9.2
Transport equipment	5,850	4,875	-975	-16.7
Other machinery and equipment	12,342	13,440	+1,098	+8.9
Miscellaneous	6,796	7,852	+1,056	+15.5
TOTAL MANUFACTURING	84,312	91,987	+7,675	+9.1

Source: calculated from Australian Bureau of Statistics data.

Table 10.2

Employment changes in manufacturing for LGAS in Western Sydney, 1984-85 to 1988-89

LGA	1984-85	1988-89	Change	Percent
Auburn	11,716	10,918	-798	-6.8
Parramatta	21,784	20,918	-866	-4.0
Holroyd	11,149	11,164	+15	negl.
Fairfield	8,787	13,559	+4,772	+54.3
Liverpool	8,558	11,087	+2,529	+29.6
Blacktown	9,805	10,848	+1,043	+10.6
Baulkham Hills	3,616	3,577	-39	-1.0
Penrith	7,287	7,490	+203	+2.8
Blue Mountains	258	517	+259	+100.0
Hawkesbury	1,352	1,909	+557	+41.2
TOTAL	84,312	91,987	+7,675	9.1

Source: calculated from Austrialian Bureau of Statistics data.

requiring lower skills and part-time or casual jobs rather than full-time employment. People losing jobs are often quite different from those gaining jobs, especially in things like age, gender and ethnicity but also in their residential location.

Global and local change and the service sector

During the 1980s, service industries began to grow rapidly in Sydney's outer suburbs although jobs in all of the major service occupations and industries remained under-represented in western LGAs compared with their locational distribution throughout the metropolitan area. While service activities have become Western Sydney's major growth sectors for employment and manufacturing has continued to shed jobs, factory-based and other blue collar work in 1991 still made up a much higher proportion of the jobs located in Western Sydney than in the rest of the metropolis. Similarly, Western Sydney's resident workforce remains much more likely to be employed in manufacturing, building and transport industries. While manufacturing workers made up less than 15 per cent of the metropolitan workforce in 1991, this proportion remained above average for Western Sydney as a whole rising to 29 per cent in Fairfield and more than 20 per cent in Blacktown, Holroyd and Liverpool (Table 10.3).

Hence, service workers remain under-represented among Western Sydney's residents. Table 10.3 shows the proportions of information workers in the workforce, defined as those in administration, community services, communications, finance and business services. While the metropolitan average in 1991 was more than 45 per cent, Western Sydney as a whole remained at 37 per cent while Blacktown, Holroyd and Liverpool showed 35 per cent and Fairfield only 27 per cent (one of the lowest proportions of resident office workers in the metropolitan area). The rapid growth of information industries, especially the finance and business services, is a direct reflection of Sydney's 'global city' functions. The location of these sectors remain heavily under-represented in Western Sydney. The recent boost of Parramatta by the State Government and private sector developers as Sydney's major suburban office centre has done relatively little to arrest this trend. Sydney's CBD, adjacent areas such as North Sydney and parts of the old Central Industrial Area, and high status suburbs along the north shore railway line, contain much of the economic activity underpinning claims about Sydney's global city status. By contrast, Western Sydney in the 1990s remains a chronic job-deficit region in terms of the number of clerical, administrative and technical jobs compared with the number of such workers resident.

Table 10.3

Western Sydney : sectoral structure of residential workforce, 1991

LGA	Total Resident Workforce: 1991	Manufacturing Workers: 1991 %	Information Workers: 1991 %
Baulkham Hills	51 739	13.7	41.2
Blacktown	75 655	20.9	35.1
Blue Mountains	25 426	9.8	50.9
Fairfield	53 778	29.3	27.4
Hawkesbury	20 448	13.8	38.9
Holroyd	30 362	20.3	35.0
Liverpool	34 602	20.6	35.7
Parramatta	50 219	17.3	41.1
Penrith	57 892	18.5	36.7
Western Sydney	400 121	19.1	37.1
North Sydney	25 732	7.4	56.1
Marrickville	29 820	17.4	42.6
Randwick	46 948	10.4	48.1

Source: calculated from Australian Bureau of Statistics, unpublished journey-to-work data 1991

Work and unemployment

Despite this under-representation of information sector jobs, continued suburbanisation of economic activity has continued to reduce Western Sydney's overall job-deficit. Census data on the journey to work from the 1991 census allow estimation of changes in the job-deficit (Table 10.4). Suburbanisation had located 81 jobs in Western Sydney for every 100 resident workers by 1991 compared with 73 in 1981 and as few as 67 in 1971. Despite this overall improvement, however, substantial job-deficits remain in localities. In the outer fringe LGA of Penrith, for example, the ratio deteriorated from 86/100 in 1971 to 64/100 in 1991, reflecting rapid growth of its residential labour force during the 1980s. Parramatta stands out as one of Sydney's major suburban employment centres with a sharp rise in its relative job-surplus. Substantial improvements in the ratio were also recorded in Holroyd and Fairfield, the former reflecting sharp falls in population growth while the latter reflects the location of new firms experiencing employment growth after 1985.

Journey-to-work data also show that, in 1991, some 64 per cent of Western Sydney's workers found jobs somewhere in the western suburbs compared with 59 per cent in 1981. Given the continued growth of the labour force, these figures show a significant increase in regional employment self-sufficiency since 1981. The improvement, however, varied significantly between localities being most noticeable in Fairfield, Baulkham Hills, and Blue Mountains. In 1991, the proportion of resident workers leaving Western Sydney each day for work remained high in Parramatta (50 per cent), Liverpool (43 per cent) and Baulkham Hills (42 per cent). Commuting has become more complex rather than simply declining as jobs have suburbanised. Information workers dominate daily flows out of the region, many travelling by public transport to the CBD or by car to northern suburban office centres and commercial estates. Movements of workers between western LGAs, mostly by car, have also increased. Hence, Western Sydney in the 1990s remains a very open labour market. Its economic activities provide waged-work for increasing proportions of its residents, although about one-third of its workers commute to other parts of the metropolis, a relatively small number to become part of the labour force for Sydney's global city functions. In addition, in 1991 just over 17 per cent of its jobs were held by people living in places like Sydney's northwestern suburbs, its inner western fringes and, especially, the Campbelltown-Camden area (a low-income, chronic job-deficit area).

The stereotyped view of Sydney's outer western suburbs as unbroken low-cost housing estates characterised by high rates of joblessness is urban mythology. This was reinforced during the 1980s by frequent media portrayals of high unemployment rates and Western Sydney as the principal locus of social problems within the city (see also Powell, 1993). At the 1991 Census, however, large areas of Western Sydney were characterised by social indicators no different from averages for other parts of Sydney suburbia. Smaller pockets showed indicators of relative affluence well above Sydney suburban averages, notably in Baulkham Hills and the Blue

241

Table 10.4
Western Sydney - ratio of jobs to resident workforce, 1971-1991

LGA	Jobs per 100 resident workers		
	1971	1981	1991
Baulkham	42	47	57
Blacktown	46	57	67
Blue Mountains	53	51	49
Fairfield	48	57	75
Hawkesbury	97	79	71
Holroyd	56	83	93
Liverpool	79	81	93
Parramatta	106	125	149
Penrith	86	69	64
WESTERN SYDNEY	67	73	81

Source: calculated from Australian Bureau of Statistics, unpublished journey-to work data, 1971, 1981 and 1991

Mountains where disproportionate numbers of (commuting) information workers reside. Indeed, analysis of both national and local media treatment of the outer suburbs in the 1990s shows, if anything, a reversal of earlier stereotyping - an attempted 'normalisation' of the West through a distinct focus on local commercial success stories such as those in Parramatta and Penrith. Yet Western Sydney's overall vulnerability to above-average unemployment rates did not fall between 1980 and 1994 and this media 'normalisation' deflects attention away from the sharply uneven impacts of economic and social change. Global economic changes such as import competition from NICs in the Asia-Pacific region, the rationalisation strategies of TNCs and the rapid growth of information rather than production sectors of the economy, are all implicated in outer suburban labour market problems but rarely treated in media representations.

Within the state planning bureaucracies, some attempts to explain the persistent above-average unemployment rates in the outer suburbs focus on the chronic job-deficits especially in suburbs where population growth rates have ensured that suburbanisation of employment has been outweighed by growth in the resident labour force. This explanation has directed urban employment policy towards expanding the local stock of jobs and towards improving the job self-sufficiency of deficit areas by attracting new investment. Other popular planning discourses have focused on the characteristics of people most vulnerable to job-loss from rationalising sectors of the economy, such as manufacturing, and show that a greater than average proportion of such people reside in the outer suburbs. According to this argument, structural economic and technological changes arising directly from globalisation since 1980 have left many people stranded in Western Sydney's labour markets as *accessible* job opportunities for which they are qualified shrink. Fast-growth sectors, such as information and financial services directly related to Sydney's international role, are concentrated heavily in other parts of the city. This explanation is often followed by calls for greater attention to technical training and re-skilling of the outer area labour force, and improvements to the 'job-readiness' of its residents.

Yet while all of these processes have affected various parts of Western Sydney during the past decade, none of the explanations deserves privileged status in accounting for local unemployment. Asymmetric power relations are principal determinants of the structure and segmentation of local labour markets (Peck, 1992, p.236). Unemployment rates in outer suburban localities increased steadily overall between the mid 1970s and mid 1980s, after which time Western Sydney began to share in the nationally falling rates (Fig 2). While the gap between Sydney and its western suburbs closed during this time, it opened sharply again with the onset of double-digit unemployment nationally in 1990. Such aggregate figures, however, hide more than they reveal. Figure 10. 2 also shows major differences between LGAs with Baulkham Hills maintaining one of the lowest unemployment rates in the metropolitan area throughout the decade while Penrith was at or below average until 1992. Consistently high rates, however, were experienced by Blacktown, Liverpool and Fairfield. By 1994, Fairfield LGA experienced one of the highest overall rates of unemployment in New South Wales.

Yet even these LGA figures bury massive differences between suburban localities in the incidence of unemployed people. Large areas of Blacktown, Holroyd and Parramatta experienced rates no higher than those in Baulkham Hills and adjacent inner western suburbs. Local labour markets in Western Sydney are strongly segmented according to such things as workers' skill-levels, and their age, gender and ethnicity. Youth unemployment captured most media attention in the western suburbs after 1980. By January 1993, Sydney's unemployment rate among 15-19 year olds stood at 21 per cent compared with 39 per cent in the Fairfield-Liverpool sub-region and 32 per cent in Blacktown, Holroyd and Parramatta. By the early 1990s, all LGAs in Western Sydney showed above average rates of unemployment among young people *including* Baulkham Hills where the vast proportion of unemployed people were under 24 years old.

While youth unemployment has united these western LGAs in a concern over future job prospects for their residents, especially given the youthful age-structure of the population, male workers over 50 also have high jobless rates by Australian standards, especially those people retrenched from manufacturing and other blue collar jobs. Further, some of the highest rates of unemployment in Western Sydney are experienced by NESB groups (females and males) with some of the highest local unemployment rates by 1994 found in suburbs with highest concentrations of recently-arrived migrants such as Fairfield and Liverpool. Problems of access to employment are reinforced by lack of English language skills (even when people can speak several other languages) especially in localities where accessible manufacturing and other blue collar jobs are disappearing. Jobs in the service sector often remain inaccessible, although they may be located in the same or adjacent LGAs, especially to retrenched manufacturing workers and young people from NESB groups. Low access to employment opportunities, for many unemployed people in the outer suburbs, has been affected strongly by their qualifications and experience which, in turn, have been influenced strongly by low access to social infrastructure such as schooling and post-school training, compared with other parts of the city. This stands out sharply in the metropolitan residential geography of university graduates and higher-status workers in the financial and business services.

Gender remains an important differentiating factor in Western Sydney's labour markets, not only in distributing workers between industries and occupations, but also shaping their experience of unemployment. Sydney's outer suburbs illustrate well the argument of Hanson and Pratt (1995) that a complex mixture of place-bound factors, gendered processes arising in the so-called domestic sphere, and workplace changes create sharply differentiated labour markets for women and men. Much larger numbers of women in Western Sydney, for example, have entered the workforce since the early 1980s as principal household breadwinners as well as 'second incomes' made necessary by rising costs of living and interest rates on mortgage repayments. By the early 1990s, overall unemployment rates among males in Western Sydney stood about two percentage points higher than for females. While this indicates fundamental changes in suburban labour markets it also disguises important trends.

Participation rates are crucial in interpreting these unemployment rates but are often ignored by the media and Western Sydney's image-makers. Four recent trends are apparent. First, participation rates for males in full-time work have declined steadily since the early 1980s especially in the inner western LGAs; indeed, most of Sydney's decline in male full-time workforce participation has arisen in its western suburbs. Second, female participation rates in full-time work have also declined in the 1990s but at a much slower rate than the decline for Sydney as a whole. Third, participation rates for both women and men in part-time and casual jobs have increased rapidly since the mid 1980s especially in Baulkham Hills, Blacktown and Penrith. Finally, by 1993 overall participation rates for females in Western Sydney remained lower than those for Sydney as a whole because of their lower participation in full-time work.

Hence, by the 1990s the lower rates of unemployment among women in Western Sydney compared with those for men still partly reflect their lower participation rates overall. Females are still discouraged from seeking waged-work if employment appropriate to their qualifications, available working hours, and daily domestic responsibilities such as child care, are not available locally or accessible by their various transport options. Lack of affordable child care facilities, for example, reduces dramatically the job-opportunities of primary care-givers (by 1995 still mostly women) especially in single parent families. Rates of hidden unemployment remain high, especially among NESB groups and, for a large number of outer suburban women, entry to waged-work has come through casual jobs with no security of tenure and low wages.

Local geography exerts a major influence over people's access to waged-work in Western Sydney. Commuting costs have a major impact on the incomes of low-wage workers concentrated in lower-cost housing areas on the metropolitan fringes. Public transport is inadequate throughout Western Sydney especially for cross-hauling between suburbs, the need for which is increasing because of the improvements in regional self-sufficiency and job-deficit ratios noted earlier. This discriminates severely against young people and women in one-car families. Use of public transport can also become a security fear for people having to cope with changed shift working hours, especially for part-time positions in factories and retail establishments. While significant numbers of female and male information workers commute to Sydney's major office centres, female commuters are skewed towards younger age-groups. Preliminary estimates suggest a high degree of localism in patterns of employment for Western Sydney's women. Access to information sources about employment opportunities, both regionally and in the wider labour market, is often relatively poor amongst the people most vulnerable to high unemployment rates. Finally, there is considerable anecdotal evidence, at least, of discrimination by prospective employers in the wider Sydney labour market against employing 'westies' given their long journeys-to-work and social stereotyping (Powell, 1993).

A model of local access to employment

Figure 10.3 summarises these local processes which determine people's access to waged-work in local labour markets. Understanding this complexity is central to local employment policy whether initiated from the 'bottom up', through local community groups, or 'top down' through State or Federal government schemes such as *Working Nation*. The stock of local jobs changes over time according to patterns of relocation of firms and rates of new firm formation. Opportunities for new business growth, and its local multiplier effects, are shown at the top of Figure 10.3. While such local growth has an obvious role to play in easing labour market problems, it is always *net* of local firm closures and patterns of retrenchment especially given the economic restructuring which is designed to accompany the globalisation strategy. Western Sydney's case demonstrates that these patterns are shaped by overall patterns of investment, for example in manufacturing and services, which are affected strongly by the Government's macro-economic polices on trade, industry, technological change and foreign investment - in short, by the turn towards international competitiveness. New employment might draw on different segments of the labour market compared with those affected by the disappearance of private sector jobs. Figure 10.3 also highlights the importance of the public sector, not only in creating or sustaining local employment in the outer suburbs but also in providing social infrastructure, where deficiencies can strongly shape labour market segmentation.

245

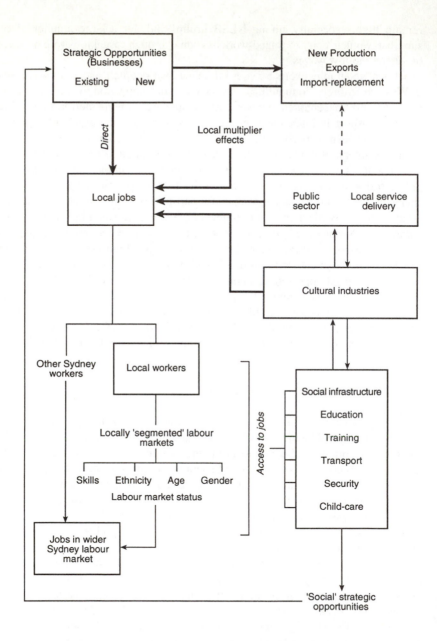

Figure 10.3 A model of access to employment opportunities in Western Sydney's local labour markets

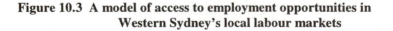

Here, the ideology of globalisation has exerted a major negative effect on Sydney's West. Tight public sector budgetary policies, at both Federal and State levels, have accompanied the globalisation strategy and have affected strongly the outer suburban areas experiencing rapid population growth. Further, the State Government invested heavily during the early 1980s in large infrastructure projects to support export processing industries, for example in the Hunter Valley. Additionally, by the mid 1980s, competition for a State share of the record foreign investment inflows (see Fagan and Webber, 1994) had switched to property development and service industries designed to capture a share of increasing international tourism from Pacific Asia. This led to major public and private sector investments in down-town redevelopment. Little of this new global and local investment in infrastructure found its way to the western suburbs which were marginalised by their lack of resource-based export industries and very limited involvement in tourist flows or global property development. Investments in infrastructure and social services lagged well behind the growth of the west, a point conceded in virtually every planning report on Western Sydney commissioned during the 1980s. The global and local coincide here to affect strongly the future competitive advantage of Western Sydney as a manufacturing or office location and of its workers in their competition for jobs.

An excess of manufacturing employment growth over job-shedding contributed significantly to net job growth in Sydney's West from the mid 1980s. More than three-fifths occurred in Fairfield where manufacturing employment grew by 54 per cent between 1985 and 1989 (Table 10.2). While large branch plants continued to shed labour, especially in the food processing and metal trades industries, Fairfield's net growth reflected small businesses manufacturing clothing, paper products, packaging and processed foods, or sub-contracting in metal fabrication. Yet over the same time-period, people's *access* to waged-work, both locally and in the wider metropolitan labour market, was determined by social, cultural and economic factors creating strong local segmentation among Fairfield's resident labour force. These local processes explain why Fairfield in the early 1990s boasted the highest growth rate of new manufacturing jobs, some of the biggest improvements in regionalisation of employment for those in work, and yet its residents experienced among the highest levels of unemployment in New South Wales.

Conclusions

Access to waged-work in outer suburban Sydney is a complex phenomenon. Persistently high unemployment cannot be interpreted simply as arising from some combination of: lack of internationally competitive industries; poor multiplier effects in the local service sector; lack of local leadership; or lack of job-readiness among the local unemployed. At the regional scale, Western Sydney does not lack these things and it is simply ideological for the state to construct its labour market problems as 'local'. While globalisation is seen to bring about a new external or outward-looking environment for urban policy, it also becomes a smokescreen for the withdrawal of the state from local and regional social investments and the winding back of planning in favour of market forces. This reconstruction of the urban and regional policy framework in the name of globalisation is happening throughout the Asia-Pacific Region (for example in South Korea).

There is nothing *intrinsically* marginal about workers or labour market conditions in the outer suburbs. Throughout the 1960s and early 1970s, LGAs in Western Sydney recorded unemployment rates either clustered at or below national rates of around 2 per cent (Fagan, 1986) - certainly well below those in Sydney's working class inner city. It is useful to reflect, however, what 'full employment' meant. National rates for males were consistently below the 2 per cent benchmark until 1975 but rates for females were nearly twice this level. Full employment was thus a highly gendered concept. The gap between male and female rates remained until the steep rises of the early 1980s. By 1993, female rates were lower than their peak in 1983 and well below male rates. The steep rises in male unemployment after 1988 were remarked upon in *Working Nation* as one of the major catalysts for government action. The much larger proportionate variation in 1967 had not been a cause for alarm.

In the 1990s, Western Sydney remains one of Australia's most productive manufacturing areas with its products sold in global, national and local markets. The characteristics which have reinforced marginality in the 1980s have not arisen by chance. Instead, they have resulted from specific kinds of economic decisions by large and small firms coupled with specific federal and state government policies implemented either in the knowledge of their uneven impacts, commonly asserting that these would be short-term 'dislocations' arising from the imperatives of globalisation, or in ignorance of their spatial consequences, commonly because affected groups have been marginalised from policy analysis, formulation and implementation.

A richer understanding of systematic marginalisation and disempowerment of people in Sydney's outer suburbs has not been well-served by some of the conventional wisdom among academics and planners about the West as a socially disadvantaged region. Nor is it served by the stereotyping that commonly employs geographical imagery of the flat, the treeless and the monotonous, which can often be sourced to 'othering' by members of an urban middle-class culture, where most of the planners and decision-makers are situated, located east of Parramatta. As Ferguson (1990, p. 12) argues, the dominant culture does not simply suppress a marginalised group (or place) but often reinvents it for its own purposes. The cleavages which, according to this paper, underpin marginalisation in the western suburbs such as gender, age-structure and ethnicity, operate throughout the metropolitan area. Yet segmentation of the labour market, and exclusion of people from access to waged-work, are both partly caused and partly reinforced by: first, the regional transport geography of the west; second, poor access to a range of social infrastructure in specific areas, often those where NESB groups and lower income-households are concentrated; third, conditions in the housing market and the public housing sector; and fourth, local demography.

Globalisation has exerted a major impact on the distribution of employment opportunities in Western Sydney, yet this paper shows that many of the positive impacts of the changes have been concentrated in other parts of the metropolis especially in inner areas and the high-status northern suburbs. Solving outer suburban labour market problems cannot be achieved simply by encouraging new firms to locate within problem localities nor by asking local councils to compete with one another in stimulating localities to 'do their own thing'. Solutions require serious attempts by *all tiers* of the state to come to grips with the social infrastructure deficiencies and more integrated planning of the social and economic environment. The question of *access* to employment considered in this paper takes us beyond job creation strategies to struggles over the provision of urban infrastructure, such as child care, health care, education and training opportunities, and affordable housing. Policies which rest on competition between places or on urging workers and localities alike to somehow lift themselves out of unemployment, simply accept by default the steam-roller metaphor of global change. Other discourses about globalisation are not only possible but required urgently.

Acknowledgements

This chapter draws from on-going research into Western Sydney's labour markets funded by a grant from the Australian Research Council. It was first presented at a conference of the International Geographical Union's Commission on the Organisation of Industrial Space: *Interdependent and Uneven Development - Global-Local Perspectives*, Seoul National University, August 1995. This was an extensive development of a preliminary version presented to a workshop on 'Social Impacts of Economic Restructuring in Australia' held by the New South Wales Geographical Society at the University of Sydney, May 1994. The author is grateful to Sherrie Cross for assistance in researching Western Sydney's labour markets, and to Mark Cole (Fairfield Council) and Graeme Larcombe for useful discussions about access to employment in the outer suburbs.

Note

1.   Unless indicated otherwise, this section is based on data collected in an on-going research project on the impacts of restructuring on local labour markets in Western Sydney funded by a Grant from the Australian Research Council.

References

Australia, (1994a), *Working Nation: Policies and Programs,* Australian Government Publishing Service, Canberra.

Australia, (1994b), *Working Nation: the White Paper on Employment and Growth,* Australian Government Publishing Service, Canberra.

Australian Geographer (1994), 'Industry, employment and regions: geographical perspectives on *Working Nation*', vol. 25, special edition.

Barnet R.J. and Cavanagh J. (1995), *Global Dreams: Imperial Corporations and the New World Order,* Touchstone, New York.

Beynon H. and Hudson R. (1991), 'More than just a storm in a pudding basin? Some reflections in and around the "localities debates" of the 1980s', paper presented to Lemnos International Seminar, 'Undefended Cities and Regions Facing the New European Order', Athens-Thessaloniki.

Burnley I. and Murphy P. (1994), *Immigration, Housing Costs and Population Dynamics in Sydney,* Bureau of Immigration, Multicultural and Population Research, Australian Government Publishing Service, Canberra.

Daly M.T. and Stimson R.J. (1992), 'Sydney: Australia's gateway and financial capital', in Blakely, E. and Stimson, R.J. (eds), *New Cities of the Pacific Rim,* Institute for Urban and Regional Development, University of California, Berkeley.

Eisenschitz A. and Gough J. (1993), *The Politics of Local Economic Policy: the Problems and Possibilities of Local Initiatives,* Macmillan, Basingstoke.

Fagan R.H. (1986), 'Sydney's west: the paradox of growth and disadvantage', in Dragovich, D. (ed.), *The Changing Face of Sydney,* NSW Geographical Society Conference Papers No. 6, pp. 70-91.

Fagan R.H. and Webber M. (1994), *Global Restructuring: the Australian Experience,* Oxford University Press, Melbourne.

Fainstein S.S. and Harloe M. (1992), 'London and New York in the contemporary world', in Fainstein S.S., Gordon I. and Harloe M. (eds), *Divided Cities: New York and London in the Contemporary World,* Blackwell, Oxford, pp. 1-28.

Fainstein S.S., Gordon I. and Harloe M. (eds) (1992), *Divided Cities: New York and London in the Contemporary World,* Blackwell, Oxford.

Ferber S. Healy C. and McAuliffe (eds) (1994), *Beasts of Suburbia: Reinterpreting Culture in Australian Suburbs,* Melbourne University Press, Melbourne.

Ferguson R. (1990), 'Introduction: invisible center', in Ferguson R., Gever M., Minh-ha T.T. and West C. (eds), *Out There: Marginalisation and Contemporary Cultures,* MIT Press, Cambridge, MA, pp. 9-18.

Gibson K. and Watson S. (1994a), 'Introduction', in Gibson K. and Watson S. (1994), *Metropolis Now: Planning and the Urban in Contemporary, Australia,* Pluto Press, Sydney, pp. 13-17.

Gibson K. and Watson S. (1994b), *Metropolis Now: Planning and the Urban in Contemporary Australia,* Pluto Press, Sydney..

251

Hanson S. and Pratt G. (1995), *Gender, Work and Space*, Routledge, London.

Harloe M. and Fainstein S. (1992) 'The divided cities', in Fainstein S., Gordon I. and Harloe M. (eds), *Divided Cities: New York and London in the Contemporary World*, Basil Blackwell, Oxford, pp. 236-268.

Healy C. (1994), 'Preface', in Ferber S., Healy C. and McAuliffe (eds) (1994), *Beasts of Suburbia: Reinterpreting Culture in Australian Suburbs*, Melbourne University Press, Melbourne.

Hodge S. (1993), 'Reading the Landscape of Western Sydney for Disadvantage', *Proceedings, Institute of Australian Geographers Conference*, Monash University.

Koc M. (1994), 'Globalization as a discourse', in Bonanno A. et al. (eds) *From Columbus to Conagra: the Globalization of Agriculture and Food*, University of Kansas Press, Lawrence.

Luke T.W. (1994), 'Placing power/siting space: the politics of global and local in the New World Order', *Environment and Planning D: Society and Space,* vol. 12, pp. 613-28.

Marcuse P. (1995), 'Glossy globalisation: unpacking a loaded discourse', in Bounds, M. (ed.), *Globalisation of the West: The Impacts of Global Restructuring on Local and Regional Development in Western Sydney*, Urban Studies Research Group, University of Western Sydney, Macarthur.

Massey D. (1993), 'Questions of locality' *Geography,* vol. 78 , pp. 142-49.

McKinsey and Co. Ltd (1994), *Business Investment and Regional Prosperity: the Challenge of Rejuvenation*, Interim Report, Department of Housing and Regional Development.

Peck J. (1992), 'Labour and agglomeration: control and flexibility in local labour markets', *Economic Geography*, vol. 68, pp. 325-47.

Porter M. (1990), *The Competitive Advantage of Nations,* The Free Press, New York.

Powell D. (1993), *Out West: Perceptions of Sydney's Western Suburbs*, Allen and Unwin, Sydney.

Rich D.C. (1995), 'The changing nature and location of work in Australia', *Geography Bulletin,* vol. 27, pp. 57-66.

Sassen S. (1994), *Cities in a World Economy*, Pine Forge, Thousand Oaks, CA.

Searle G.H. (1996), *Sydney as a Global City*, New South Wales Department of Urban Affairs and Planning, Sydney.

Taskforce on Regional Development (1993), *Developing Australia: A Regional Perspective*, Australian Government Publishing Service, Canberra.

11 Localisation and globalisation tendencies in the social control and regulation of labour

Andrew E.G. Jonas

Introduction

In recent years, writers on the labour process have examined the ways in which social relations outside the immediate work environment influence and shape employment relations at the point of production (Burawoy, 1985; Littler, 1982; Maguire, 1988; Warde, 1988). According to these writers, employment relations are shaped by locality-specific ideological and political developments. In order to understand this 'locality politics of production' it is necessary to consider capitalist relations in any given context as constituting a 'social totality' of production, reproduction, and consumption relationships. This in turn entails a shift in emphasis, from the politics of *work*places (i.e. labour control practices internal to firms and industries) to that of work*places* (labour control practices and systems of labour regulation transcending the boundaries of firms, industries and labour markets within particular localities) (Peck 1996).

At the same time as some researchers have begun to examine the contribution of locality-specific social relations to the internal politics of production, it has become evident to others that the wider (global) context of labour regulation and control is undergoing a profound change as capitalism restructures from a mass production to flexible regime of accumulation (Storper and Scott, 1989), and as firms and industries develop global production strategies (Dicken, 1992). The social regulation and reproduction of labour is increasingly seen to be influenced, if not determined, by globalisation processes which cut across the boundaries of localities and labour markets. These processes include the movement of manufacturing capital from high to low wage countries, the emergence of a New International Division of Labour, the subordination of labour to the imperatives of flexible production, the intensification of inter-locality competition for inward investment, and the global crisis of the Fordist-Keynesian welfare state (Lipietz, 1987; Jessop, 1990, 1994). For some, these processes have ushered in a new era of labour control in which globally mobile capital has forced labour to retreat into its communities and defend its place-based interests (Burawoy, 1985; Herod, 1991, 1994). The new 'global politics of production' has denied labour the opportunity of mobilising around progressive social contracts and alternative production strategies (Fitzgerald, 1991; Tickell and Peck, 1995).

These developments in the literature capture two contrasting approaches to conceptualising the geography of labour control. On the one hand, the recognition that labour control involves more than simply the exercise of managerial control at the point of production has sharpened awareness of locality-specific relations and conditions of labour reproduction, suggesting that the local scale offers the most satisfactory vantage point from which to examine current changes. On the other hand, the idea that what happens to workers in any particular locality has more to do with economic globalisation than political processes within localities implies that the global scale is the more appropriate level for understanding the new labour control imperatives of flexible production.

Apparently, there is no meeting ground between these two perspectives, and no prospect of developing middle level abstractions with which to conceptualise the 'new' geographies of labour control.

In this chapter, I argue that neither the globalisation perspective nor the locality approach provide entirely satisfactory vantage points from which to examine contemporary developments in the geography of labour control. Rather each approach highlights elements of what are intrinsically *uneven and changing geographies* of labour regulation and control. While globalisation processes remain an important part of the changing wider context of labour regulation, and while restructuring through space must always be seen as a potential threat to any local labour control regime, the development of locality-specific labour control practices and systems of labour regulation remains a structurally-given feature of production in an unevenly developing economy. The chapter therefore discusses three 'types' of local labour control regime and concludes with a review of the implications of recent developments for the changing social and geographical basis of labour resistance.

Globalisation tendencies: spatial restructuring and labour control

Recent research has paid increasing attention to the relationship between labour control and the reorganisation of production structures across space (Berberoglu, 1993). It has also been suggested that geography - in the sense of the existence of a necessary tension between mobility and immobility - lies at the heart of a globally hegemonic labour control regime.

An historical tendency for production structures to avoid localities where workers are organised and resistant to management has been noted by several scholars. Gordon (1984) cogently argued that the suburbanisation of industry in the United States was a direct outcome of labour control problems facing corporations. The American metropolis has evolved in distinct stages which correspond more or less to each development in the labour process. In a similar vein, the avoidance of organised (union) labour has been seen to be responsible for the wholesale deindustrialisation of the North East region of the United States (Bluestone and Harrison, 1982). Thus the industrialisation of the South was premised on the ability of relocating corporations to find reserves of tractable labour in 'right-to-work' states. Deindustrialisation, regional restructuring and the emergence of a New International Division of Labour have been viewed as geographical aspects of the crisis of American Fordism, both as a system of production and a means of control, which gained momentum during the 1970s and 1980s (Lipietz, 1987).

More generally, it has been argued that the inter-regional mobility of productive capital forms a necessary pre-condition for the establishment of new regimes of accumulation. In the current context, a regime of flexible accumulation is most likely to have taken hold in regions and localities where Fordist labour practices have not become entrenched:

> ... as flexible production began its historical ascent in North America and Western Europe, it tended to flourish most actively at places where the social conditions built up in Fordist industrial regions either could be avoided or were not present...(P)re-exiting clusters of flexible producers (in central business district areas or in traditional craft communities) have been revitalized and have grown very rapidly of late..and a series of new industrial spaces has come into being in the various Sunbelts, "third development zones," and suburban peripheries of the advanced capitalist countries. In all of these production locales, the social and/or geographical distance from the old foci of mass production is great. (Storper and Scott, 1989, p. 28)

And the reason why 'old' production centers are not conducive to the 'new' regime of accumulation is because:

> ...(the) extended historical experience of industrial production and work (in particular localities) tends to have as its correlate the rigidification of labour relations and the unionization of large segments of the working class. (Storper and Scott, 1989, p. 27)

There is implicit in this argument a tendency for 'new' firms and industries to avoid concentrations of unionised labour, a tendency which has become more pronounced in recent years because of the growing imperative to organise production strategies around flexible labour markets and employment relations.

Along with these developments in the regulatory framework of production, it has been claimed that productive capital has become more mobile vis-à-vis labour, which for the most part remains bound to particular places (Storper and Walker, 1983, 1989). This mobility forms the basis of a new 'global politics of production' in which the mere threat of relocation is used as a form of political leverage by employers over employees in the determination of wage and benefit packages, contracts, investment strategies, and production goals. Corporations are thus able to extract concessions from workers and communities because' ... differential capital investment and the potential for capital mobility have given corporations great political and economic leverage over community politicians and workers' (Herod, 1991). For their part, workers have resorted to defending their interests in particular places and production strategies, and in doing so have found themselves having to compete with workers in other places (often within the same production structures) to keep plants operating, jobs secure, and communities stable For some researchers, then, the age of globally-mobile capital has ushered in a new despotic system of labour control. 'Hegemonic despotism' is characterised less by the search for consensus in the workplace and more by the fear of plant closures and job losses resulting from capital flight, disinvestment and the transfer of operations (Burawoy, 1985).

Three caveats to the above claims are worth mentioning. Firstly, although plant closures and deindustrialisation have decimated the ranks of the 'mass collective worker', the decline of unions and their power to challenge management practices has been uneven across space. In short, some localities remain more resistant to spatial restructuring and capital mobility than others (Jonas, 1995), suggesting that 'hegemonic despotism' is far from being hegemonic practice in particular localities.

Secondly, the negotiation of contracts on a plant level rather than national basis has enabled corporations to play one locality off against another. Nevertheless, workers have been able to build a solid basis of community support in localities where they have greater influence over plant level production strategies.

Although the relationship between production and community has clearly changed, it is important to recognise that even at the height of mass production the level of community dependence of capital was quite high.

Thirdly, while capital in general has become more mobile, specific sectors and companies are often limited in their spatial restructuring strategies by virtue of the local production systems into which they are locked. For example, despite the recent tendency for dispersal away from its traditional production centers, the Midwest auto industry has witnessed a remarkable process of re-agglomeration at the regional scale. On the one hand, the decentralisation of Japanese assembly plants *has* been driven by the desire to reconstitute management-labour relations away from Fordist centres. On the other hand, a process of re-agglomeration has been encouraged by the need for assembly plants to retain control over their transplant supplier operations within a just-in-time production system (Mair, Florida and Kenney, 1988). In short, once a production regime becomes established in a given territorial setting, there are limits to spatial restructuring contained within its production structure. There comes a point at which industries and firms are unable to resolve their labour control problems by relocating but rather must address them by remaining in place.

Localisation tendencies: reciprocal labour control practices

An important corollary to globalisation tendencies in the control of labour is therefore the *geographically-embedded* nature of production and labour control. In order for accumulation to occur, labour power must be combined with means of production at particular locations. This means that capitalists must continue to exercise control *in situ,* not only over the social and material aspects of the labour process, but also over the place-specific conditions under which labour is reproduced and becomes available in the labour market.

Employers self-consciously came to recognise the importance of control over the wider (locality) conditions of labour reproduction towards the latter part of the nineteenth century. This was a time when the contradictions of factory paternalism were becoming increasingly evident. Paternalism failed not so much because of the inability of employers to control conditions internal to the workplace but rather because of a failure to generate reciprocal (workplace-community) relations external to it, and hence conducive to accumulation (Melling, 1992). Industrialisation and urbanisation brought together workers who, despite coming from a variety of ethnic and cultural backgrounds, shared in common the experience of low pay, long hours and unsafe working conditions (Hareven, 1982; Zunz, 1982). Cities afforded new opportunities and sites for workers to organise and challenge the despotic conditions of factory production. Strikes and labour unrest in the latter decades of the nineteenth century marked a general breakdown in means of social control in industrial paternalistic capitalism (Gutman, 1976).

This period also marked the transition between machinofacture and mass production. With the introduction of assembly-line production, the development of reciprocal (workplace-community) relations became an increasingly necessary means by which employers were able to enforce the terms of the labour contract. Factory owners found it to their advantage to foster cultural values in the wider community conducive to (mass) production (Nash, 1987). The transition from paternalistic to welfaristic form of social control was driven as much by a desire to reassert control outside the factory as within it: 'It is more satisfactory to locate the provision of social benefits in the employers' continuing search for compliance in the workplace *and the locality*' (Melling, 1992, p. 118; emphasis added). The foundations of Fordism, then, did not simply lie in the application of Taylorist methods of scientific management to the labour process but rather also in the development of an holistic approach to labour control (based around corporate welfarism) which allowed the underlying coherence between mass production and mass consumption to be realised. In this respect, Henry Ford was one of the first industrialists to recognize the contribution of the 'totality of social relations of production' to the effective integration of workers into the labour process (Gartman, 1993; Harvey, 1989).[1]

It is only comparatively recently that industrial relations scholars have recognised the strategic inter-relationships between locality, the workplace, and the wider conditions of labour control. Initial research in this vein focused on the role of the industrial and community milieux in shaping working class consciousness and deference (Lockwood, 1966; Newby, 1977; Norris, 1978). More recent work has examined the functioning of extended internal labour markets and the development of local loyalty systems in the control and integration of workers into the labour process (Granovetter and Tilly, 1988; Littler, 1982; Manwaring, 1984). Scholars here have highlighted the false nature of the distinction between work and community in the control of labour (Maguire, 1988). Thus Maguire's '..analysis of the local context and the informal methods of recruitment emphasised the interpenetration of 'external relations' on the one hand, and control elements arising from the labour process and the organisation of work on the other' (Maguire, 1988, p. 83).

Extending the ideas of Burawoy (1979; 1985) and Massey (1984), Warde (1988; 1989) has offered a general framework for understanding the role of labour control in the development of a locality-specific 'politics of production'. He suggests that local 'factory regimes' are influenced by the variety of social, political and cultural institutions and ideologies around which workers organise their daily lives and loyalty systems. Thus over time each locality develops its own distinctive 'politics of production' based around local patterns of production, consumption and reproduction. Geographers have added their own insights into this emerging literature on the relationship between local labour control practices, local workplace cultures, and processes of economic adjustment within localities (Jonas, 1992, 1996; Hanson and Pratt, 1992, 1995; Peck, 1992, 1996).

More generally this research has considered the social conditions and institutions which regulate the functioning of local labour markets (Peck, 1996). Under Fordism, as localities became integrated into national and international production and exchange systems, the reproduction of labour became separated in time and space from the immediate domain of production, and it became increasingly necessary to develop wider social institutions for regulating labour markets (and hence the supply of labour at the point of production). It was (and is) a contingent matter that these wider systems of labour-market regulation came under the control of the state. Nevertheless, the Keynesian welfare state represented a deliberate attempt to centralise key elements of labour regulation in Fordism, including training, education, collective bargaining and income support, with a view to meeting the needs of national industries. With the onset of the crisis of Fordism, however, states began to experiment with more decentralised systems of education provision, job training, wage bargaining and income support. Arguably these decentralised systems may yet constitute the basis of a new system of labour regulation - the 'Schumpeterian workfare state' (Jessop, 1994) - in which there has been a deliberate strategy of localising labour in order to foster a regime of labour control conducive to global accumulation (Peck, 1994b). However, even during the Fordist era there was a great deal of national and local variation in regulatory frameworks. In the United States, for example, a number of industrial sectors (and national unions) had decentralised systems of contract negotiation and bargaining (Davis, 1986), and states and localities retained significant powers to regulate the conditions of labour reproduction. The fact that programmes and policies, such as workers compensation and education, resisted federalisation during the New Deal is testimony to the uneven spatial development of the mode of social regulation in US Fordism.[2]

Much also has been made about the crisis of mass production and what that entails for the revival of craft-based forms of production and control. Some commentators suggest that neo-paternalistic forms of labour control are associated with the revival of industrial districts (Piore and Sabel, 1984), while others claim that the 'new workplace cultures' of post-Fordism are based around decentralised management systems which allow for employee participation in the management, ownership and control of production (Appelbaum and Batt, 1994). Industrial districts and the rise of networks of small, flexible firms are believed to herald a new era of global competition in which close inter-firm networks develop among competing producers (Best, 1990; Piore and Sabel, 1984).

It is still far from clear, however, as to what changes in the organisational basis of production and control entail for the 'locality politics of production'. Are these 'new workplace cultures' based around the strengthening of external labour market ties and the further embedding of local relations of reciprocity?[3] Or are the pressures of global competition weakening such ties and undermining such relations? To what extent can we talk about *new* social and geographical bases of labour control in which localisation rather than globalisation tendencies are dominant? Rather than providing a definitive set of answers to these questions, I attempt in this chapter to develop a framework which acknowledges a combined role for localisation and globalisation tendencies in the shaping of a 'new locality politics of production' and 'new geographies' of labour control (recognising of course that elements of 'old geographies' are contained in the 'new'). At the core of this framework is the idea of a 'local labour control regime'.

Localisation and globalisation tendencies: local labour control regimes

The idea of a local labour control regime is grounded in the regulation approach of political economy, in particular its reading of the relationships between labour control and the wider social basis for accumulation (Gordon, Edwards and Reich, 1982).[4] So-called 'regulation theory' focuses on the causal and contingent relationships which exist between the micro-management of the labour process and the macro-regulation of the economy (Aglietta, 1976). Essentially, the approach argues that stable periods of economic growth, or 'regimes of accumulation,' emerge from struggles around, on the one hand, the social and technical organization of the labour process and, on the other, the wider social conditions of wage determination, consumption, and labour reproduction. Each regime of accumulation thus has its own corresponding 'mode of social regulation' (MSR). The MSR is comprised of an ensemble of institutions, policies, practices and spatial patterns which structurally enable the circuit of capital to be completed, thereby melding together production strategies and consumption practices into (a) dominant accumulation strateg(ies/y) (Jessop, 1990).

Although the foci of regulation theorists have become quite diverse, one issue that continues to be of concern is the precise causal connection(s) between locality-based labour control practices and the wider mode(s) of social regulation within which those practices are located. Is it possible, for example, to talk in terms of distinctive 'local modes of social regulation' as constituting the seedbeds of larger-scale (national or global) modes of regulation (Peck and Tickell, 1992)? If it is possible, what role does labour control play in such 'local modes of social regulation'? If not, can there still be structurally coherent labour market processes operating at a time when wider regulatory frameworks are in such disorder and local labour markets are under increasing pressures from globalisation?

The regulation approach has not provided satisfactory answers to these questions (and in many respects has not even begun to ask the questions!). However, a way forward might be found in linking the basic principles of regulationist thinking to the idea of uneven development. Some of this work has already begun (Florida and Jonas, 1991; Peck and Tickell, 1992; Tickell and Peck, 1995). In the context of labour control, any new approach along these lines would need to highlight three sets of local-global relationships.

First, although regulation theory clearly recognises the importance of establishing causal links between the micro-regulation of the labour process and macro-management of the economy, it has mainly tried to locate those links in regulatory processes operating at the national scale. In doing so, regulation theory misses at least two other crucial scale dimensions of regulatory change. On the one hand, it overlooks the local scale where, as I have suggested, reciprocal relationships may already have developed around local labour markets. On the other hand, regulation theory remains agnostic about global-scale regulatory processes and, in particular, the regulatory implications of the New International Division of Labour. In short, the national scale is not necessarily *the* appropriate level at which to identify new regulatory frameworks conducive to the meshing of production and consumption (Tickell and Peck, 1995).[5]

Secondly, the emphasis on uneven development allows one to recognise, not simply national variations in labour market conditions, but also the role of sub-national variations in conditions governing the social regulation of labour. While to be sure there are considerable national variations in wage conditions and modes of labour reproduction, these variations can conceal significant sub-national variations, notably the operation of locality-specific labour market processes. These sub-national variations might account for why firms and industries have chosen not to restructure through space in response to the failure of national or local regulatory frameworks. Moreover, such variations might explain why 'global Fordism' itself has been a geographically selective process (Lipietz, 1987).

Thirdly, assuming that it is possible to identify coherent 'local modes of social regulation,' it is not entirely clear whether these form in response to the uneven restructuring of capital across space or are generated *in situ* from the development of a localised production structure. While those writing from a spatial divisions of labour perspective would tend to emphasise the former process, recent research on industrial districts and vertically-disintegrated production structures would suggest that the latter generative process is dominant. In some respects, both types of narrative are 'right', and in other respects both are 'wrong'. Perhaps then a more important issue is whether such frameworks, if they do exist, have the capacity to endure, especially in a rapidly globalising context. Tickell and Peck (1995) are currently doubtful about the attainability of sustainable local social contracts between labour and capital; but I would suggest that it is possible to identify the generic properties of already existing local labour control regimes.

The advantage of the regulation approach is the ability to recognise at an abstract level the underlying unity and structural coherence of production, consumption, and labour reproduction. At a concrete level, whether that coherence is achieved through more localised or conversely more globalised regulatory frameworks is a contingent matter. And just as it is unlikely that any post-Fordist regulatory fix can be imposed 'from above' on to localities in a pure form, so also is it unlikely that whatever regulatory fix comes 'from below' will become generalised across space in the same form as it originated in place. The seedbeds of local (and global) modes of social regulation are always and necessarily to be found in labour control practices which are to a greater or lesser degree locality-specific.

The idea of a 'local labor control regime' takes as its point of departure this necessary tension between globalisation and localisation in the social regulation and control of labour. On the one hand, it encapsulates the reciprocal labour-market practices which tend to bind together employers and employers in any given localised production, consumption and reproduction setting. One the other hand, it recognises that the endurance of these practices is contingent upon developments in the wider economy and regulatory structure. Thus the long term reproduction of these local relations of reciprocity is problematic.

Any threat to the structural integrity of a local labor control regime may encourage locally-based companies and industries to intensify or reorganise local labour market reciprocities, to avoid labour control problems by restructuring through space, or to transplant production structures elsewhere. Local labour control regimes can provide the seedbeds of, or form an important local element within a, wider (i.e. *ex situ*) system of social regulation. Conversely, the failure to reproduce *in situ* local labour-market reciprocities can create a local regulatory crisis, leading to the transformation of the local labour control regime.

A local labour control regime is identified both by the uniqueness of its constituent labour control practices and by the extent to which those practices have endured in a particular geographical setting. In certain settings, labour control practices have endured sufficiently long to bind together ensembles of firms, industries, wororrs, etc. into a relatively cohesive territorial structure, such that any threat to the territorial integrity of that structure has required some form of self-conscious exercise of local regulatory capacity. The critical issue then becomes identifying the nature of that regulatory capacity. Rather than fitting regimes into generic 'types' (since, by definition, each regime is going to differ in terms of its dominant characteristics), I prefer to define a regime's capacity in terms of the 'function' (i.e. production strategy) performed by the control of labour.[6] The three case studies presented here therefore have been chosen to exemplify the functional capacities of different 'types' of local labour control regime.

Case studies

The choice of these particular examples is justified on the following grounds. First, they represent three very different localities: an 'old industrial space'; a 'new industrial space'; and an 'industrialising space'. They therefore encapsulate a representative range of production strategies and labour market processes. Second, they have endured sufficiently long as territorial production ensembles to suggest there is some degree of structural coherence operating at the local or regional scale. Third, each locality has recently experienced globalisation pressures, albeit the nature of these pressures and the responses to them have been locally unique. Fourth, the responses in all cases have seen a tendency towards the increasing localisation of capital, particularly as far as its level of involvement in activities relating to the control and reproduction of labour is concerned. Fifth and finally, while I have presented each locality as an individual case study, it should be remembered that the experiences of these localities have not occurred in isolation from each other or indeed from the experiences of other, similar industrial spaces (and local labour control regimes) in North America and elsewhere.

Local control for production: Worcester, Massachusetts

With spatial restructuring and deindustrialisation dominating the agenda of many localities in the United States manufacturing belt in recent years, it is instructive to consider the fortunes of one industrial centre in the region whose distinctive production structures and labour market processes can be traced as far back as pre-Fordist era. Worcester (central city population 165,000 in a metropolitan area of over 400,000) has for well over a century been one of Massachusetts' largest and most important industrial centres, second only to Boston in terms of population size. Although not immune to deindustrialisation and increasing external control, production in Worcester has been built around an industrial culture which has emphasised the virtues of local control and of the active participation of local employers in the social and political fabric of the community. This unique culture is embedded in the labour control practices of Worcester-based manufacturing companies, which can be traced to the transition period between craft and mass production.

Worcester's early industrial development was based upon textile, corset and wire manufacturing. As the local economy grew, it diversified with the addition of glass, metalworking, paper and envelope, abrasives, and castings industries. Many of these industries were craft-based enterprises founded by immigrants from countries in Northern Europe, including the United Kingdom, France, Germany and Sweden. As these enterprises expanded, mass production methods and Taylorist principles of scientific management were gradually introduced, and labour processes became increasingly segmented by skill, gender and ethnicity. Nevertheless, Worcester-based companies resisted the wave of external consolidations which swept through American businesses towards the end of the century, preferring instead to merge with local companies. This pattern of local ownership and control characterised the Worcester economy until the 1960s, and helped to sustain a culture of anti-unionism amongst the local ruling elite.

> Sustained local ownership of the city's largest companies advanced the anti-union cause and, of course, thickened the "complex web of informal connections" that fostered anti-union solidarity among Worcester industrialists. (Hanson and Pratt, 1995, p. 50)

In terms of the widespread adoption of an anti-union political culture, some see the labour unrest of 1915 as a pivotal event in Worcester's history (Rosenzweig, 1983; Cheape, 1985). The de-skilling and standardisation of the labour process in first decade of this century led to a demand for semi-skilled labour, which was largely met by immigrants from southern and eastern Europe. By the outset of World War I in Europe, Worcester-based manufacturing plants were increasingly segregated on the lines of skill, ethnicity and gender. At the core of the labour market were skilled craft metal and abrasives workers from northern Europe who by and large enjoyed reasonable pay and working

263

conditions, and held more privileged occupational standings. At the periphery of the labour market were unskilled and semi-skilled immigrant workers from southern and eastern Europe, who enjoyed few on-the-job privileges, and experienced unstable employment prospects, and unsafe working conditions. Patterns of ethnic, gender and occupational segregation within the local social structure were reinforced by the paternalistic labour control practices of Worcester industrialists which explicitly took into account the different backgrounds of employees (Jonas, 1992).

The situation for unskilled and semi-skilled workers was not helped by low rates of unionisation in Worcester. In 1910, only 10.7 per cent of the Worcester labour force was unionised. This compared to 25.5 per cent in Lynn, Massachusetts, and 21.2 per cent in Boston. Nevertheless working Worcesterites did organise and attempt to confront the paternalistic practices of their employers. Whereas in other US cities undergoing rapid industrialisation, workers developed political consciousness through union membership or the political machine, workers in Worcester orientated themselves around the cultural worlds of the city's growing and diverse ethnic neighbourhoods, where they shared in common their experiences of low pay, long hours and unsafe working conditions (Rosenzweig, 1983).

In 1915, tensions among the working class population encouraged to organised resistance to the labour control practices of Worcester companies. The International Association of Machinists launched a strike in a number of local machine and metal factories, demanding pay increases and a shorter work day. The response of the factory owners was harsh and decisive. They forbade their work forces from affiliating with the union, deployed spies and, through the well-financed local Metal Trades Association, the media and local politicians, orchestrated a campaign of anti-union propaganda, which included the blacklisting of labour organisers and the banning of courses on trades unions in the local school system. Within a matter of weeks, they had regained control of their work forces and the strike was crushed.

The 1915 strike nevertheless exposed industrialists to the limitations of owner paternalism, and encouraged them to adopt a welfaristic approach to labour control. Unlike the bureaucratic practices of large vertically integrated enterprises elsewhere, the corporate welfarist culture that evolved in Worcester was motivated by the desire to maintain and assert local control of production. Local employers became major donors to local charities, organised community-wide holidays and picnics, supported local schools, constructed health and recreational facilities for their employees, and generally become more directly involved in the everyday activities of the community. These practices fostered strong industry-community ties, and saw the development of extensive social networks linking workplaces and residential neighbourhoods throughout the city. Local companies also ran candidates in local elections. Prominent positions in city hall were occupied by senior personnel from companies like Norton and Wyman Gordon. An industrial patronage system developed whereby in exchange for ease of access to local sites and city services, Worcester firms agreed to expand production locally and to contract with local suppliers and service providers, thereby contributing to the local dependence of what in some cases had become very large organisations (Jonas, 1992).

The historical legacy of the welfaristic labour control practices of Worcester companies is manifested in three ways: the perpetuation of a local culture of anti-unionism; the dominance of reciprocal labour recruitment practices; and strong resistance to external control from within the local civic and industrial leadership.

Firstly, then, Worcester has been, and remains to this day, a 'non-union town'. Although at 34 per cent the rate of unionization in Worcester in 1950 did exceed that of the United States (31 per cent), by 1990 it was down to 14 per cent, or 2 per cent below the national rate, and roughly the same rate as it had been in 1920. Despite ongoing attempts by national and international unions to organise the local work force, many if not most of Worcester-based manufacturing (and service) firms have no union affiliation. A survey of 149 Worcester-area firms revealed that only 10 per cent of the surveyed firms employed unionised workers (Hanson and Pratt, 1992). Over the years, attempts to organise workers at local companies have faced resistance, not so much from owners and senior management, as from the workers themselves. In the case of Norton Company, for example, employees have continued to honour a pledge signed after the 1915 strike not to 'knowingly do anything contrary to the best interests of the Norton Grinding Company' (Cohen, 1988, p. 163) and have resisted every opportunity since then to form or join a union. One Norton employee remarked to the effect that even when the company (undergoing restructuring during the early 1980s) was laying off hundreds of its Worcester employees the thought of joining a union never entered his mind.[7]

Secondly, the recruitment and hiring practices of Worcester firms have engendered a close network of ties between the city's workplaces and residential spaces. Inter-generational recruitment practices remain commonplace, and firms focus their recruitment campaigns on specific neighbourhoods within the metropolitan area. Not only have these practices contributed to patterns of occupational and gender segregation within the workplace, they have intensified the dependence of Worcester firms on local labour markets within the metropolitan area to the point where they are extremely reluctant to relocate :

> [Worcester] employers are keenly aware of the characteristics (skills, costs, gender) of the locally available labor implied by a given location. Moreover, once settled in a particular location in order to tap that labor supply, most employers develop close ties to that highly localized labor force - ties that they are reluctant to disrupt even when the need for more space demands a change of location. (Hanson and Pratt, 1992, p. 391)

When Worcester firms do need to move, because of such 'dynamic and reciprocal labour-market dependencies' they tend to relocate within the metropolitan area rather than outside it so as not to disrupt the place-bound ties of workers, many of which orientate their job-search behaviour around their locations of residence (Hanson and Pratt, 1992, p. 392).

Thirdly, the economic and political leadership of Worcester remains suspicious of, even resistant to, the increasing external control of the local economy. Although signs of increasing external ownership and control in New England were already apparent in the 1960s (Dicken, 1976), it is fair to say that the Worcester industrial and political leadership was slow to respond to the inward investment opportunities which were being created. The city's paternalistic employers feared that inward investment would encourage unionisation of their work forces and put upward pressure on wages. These concerns were compounded by the fear that, as the city tried to meet the needs of incoming investors, local manufacturers would lose their influence in local political and economic decision making. Indeed, by 1986 local manufacturers represented only 19 per cent of the directorship of the Worcester Area Chamber of Commerce, compared to 48 per cent in 1962.

As a result of the acquisition of prominent local companies, such as the Paul Revere Insurance Company and The Worcester Telegram and Gazette Inc., local organisations have been forced to change their outlook on local economic development. As one local publication put it:

> Perhaps because Worcester's business climate has long been pegged "parochial," outsiders have all but overlooked the city as a place to expand....In recent years, though, Worcester has lost control of much of its industrial base...Ironically, the only way the city can achieve significant economic growth now is through an influx of outside concerns. (Kowal, 1990, p. 6)

In response to the 'new economic reality' of inward investment, there was a major drive in the 1980s to recruit new personnel to the city's business leadership, to widen the scope of its economic development activities, and to attract outside investors to Worcester. As a result of these actions, the city was able to attract BASF Bioresearch Corporation (a Germany-based conglomerate) to a new research park located near the University of Massachusetts Medical Center in Worcester.

However, debates within the local leadership about the merits of having a foreign company like BASF in Worcester showed that local suspicions about external ownership continued to be harboured. Feelings came to a head in 1990, when a British conglomerate, BTR, made a bid to acquire Norton Company, the 'jewel in Worcester's crown'. Norton Company, a manufacturer of abrasive products and the city's largest private employer, was the very embodiment of the paternalistic/welfaristic labour control practices of Worcester-based companies. In

1990, it was a multinational enterprise emerging from a major internal restructuring in response to a shrinking domestic market for abrasive products. It had diversified, sold off several subsidiaries, and created large cash reserves, making itself an attractive target for a takeover.

There was an outpouring of community sentiment against the acquisition bid. For many, including Norton employees, the company represented the final bastion of resistance against external control and manifested all the positive qualities of community involvement. The anti-takeover movement organised by hourly employees was widely supported at the city and state levels. Legislation was hurried through the Massachusetts legislature which resulted in a tightening of the state's anti-takeover laws. But the action taken came at a price for the Worcester industrial leadership, for it meant that the company could be saved only through a friendly merger with a French conglomerate, Compagnie Saint-Gobain. Many of the concerns harboured about external control have since been justified. Although Norton Company has invested $US60 million in a new local production facility, the new management has showed signs of reducing its level of community support and relations with employees have become strained, resulting in recent drives to unionise the work force. If events at Norton Company are representative of forces at large in Worcester, then there are signs that the local labour control regime which has been built around local control for production is under threat of collapsing.

Control for innovation: semiconductor firms in Silicon Valley

The regional agglomeration of semiconductor and computer firms in Silicon Valley, Santa Clara County, California, is widely recognised for its innovative capacity and unique culture of competition (Angel, 1994; Saxenian, 1994, 1996). Since its founding in the 1950s, the agglomeration has continued to spawn new firms, demonstrating remarkable resilience in the face of intense global competition. While rival growth centres in the United States such as Boston's Route 128 industrial complex have experienced profound problems, Silicon Valley is in the midst of the most rapid growth phase in its history and continues to generate new ventures and products (Saxenian, 1996). If Worcester exemplifies the resilience of a local labour control regime in an 'old industrial space,' then Silicon Valley provides evidence of labour market processes operating in a 'new industrial space' which have also shown a capacity for resilience in the face of globalisation.

Explanations for Silicon Valley's success have tended to emphasize the qualities of its elite work force, especially those of its pool of highly qualified engineers, the flexibility of its firms and governing institutions to adjust to new product and process technologies, and external economies deriving from industrial localisation (Angel, 1994; Saxenian, 1994, 1996). Comparatively less attention has been paid to the role of labourmarket processes and labour control strategies in the endurance of the industrial complex. Although Silicon Valley

267

meets most of the labour requirements of a 'new industrial space' (Scott, 1988b; Storper and Scott, 1989), notably the absence of rigid Fordist employment relations and an emphasis on decentralised management practices which facilitate inter-firm exchanges (Saxenian, 1996), little attempt has been made to consider the labour market processes operating within the complex from the vantage point of labour control in its 'social totality'.

A review of the relevant literature suggests that labour market processes operating in Silicon Valley are geared towards maintaining the coherence of the industrial agglomeration as a locus of innovation (Angel, 1994; Saxenian, 1994). Three sets of processes appear to be central to this 'control for innovation' function: locational clustering; facilitating inter-firm mobility; and minimising external competition for labour.

The locational clustering of new semiconductor firms in Silicon Valley has been central to the region's success as a centre of innovation. Most of the major producers, merchants and users of semiconductors in the USA are located in Santa Clara County, enabling these firms to realise economies of agglomeration in labour-market processes (Angel, 1994). Firms recruit skilled workers from the regional labour market, thereby avoiding the costs of developing their own in-house training programmes. Ninety per cent of technicians and process engineers are recruited in this fashion by Silicon Valley firms. This recruitment strategy is especially important for start-up firms, which need to be able to assemble research teams quickly from the local pool of skilled labour. In the case of process engineers, Silicon Valley firms meet over 85 per cent of their recruitment needs from within the regional labour market compared with 33 per cent for firms located outside Silicon Valley (Angel, 1994). Even the majority of unskilled labour is recruited locally. In the case of fabrication workers, 100 per cent recruitment from the local labor market is common practice among local semiconductor firms. This compares with semiconductor firms elsewhere in the USA, which recruit mainly from outside the regional labour market.

Locational clustering alone does not explain Silicon Valley's success as a centre of innovation. Rather it is how the process of sectoral locational clustering has enabled inter-firm social networks to develop within the industrial complex which is important (Saxenian, 1996). Such networks have produced (and reproduced) a localised culture of collective learning and entrepreneurial innovation (Angel, 1994). Inter-firm networks have facilitated the rapid communication and exchange of ideas and information about new technologies, products and labour processes. Thus the high inter-firm mobility of core workers (engineers and managers) has become critical to the reproduction of a culture of innovation within the industrial complex. This continuous ability to generate new spin-offs is largely responsible for the diversification of Silicon Valley's economic base from a focus on semiconductor production in the 1970s to a wide range of activities specialising in computer software and hardware.

In order for Silicon Valley to reproduce itself as a centre of innovation, it is more important for core workers to remain within the industrial complex than it is for them to remain loyal to any particular firm in Silicon Valley. The close geographical proximity of firms has facilitated the types of inter-firm mobility and entrepreneurial exchanges central to the region's innovation capacity, even as external labour market pressures have grown. These pressures have resulted from the recruitment practices of rival growth centres in the South and West. Silicon Valley firms have responded by offering wage and other incentives to 'buy' worker loyalty. Such practices were quite commonplace in the late 1970s and early 1980s, but have also contributed to the emergence labour market diseconomies of agglomeration (see below).

For a while, the globalisation of semiconductor production did threaten the integrity of localised labour market processes in Silicon Valley. Beginning in the 1970s, firms dealing with mature products began to relocate production outside Silicon Valley to places offering lower land and labour costs. Advanced manufacturing went to regional growth centers in the South and West of the United States, and mass assembly operations were relocated to offshore locations, including parts of South East Asia (Scott and Angel, 1988). Silicon Valley semiconductor firms began to develop global production strategies while retaining locally-based control and research functions.

This process of regional restructuring came in response to emerging overseas competition in the semiconductor industry. But in the early 1980s resistance to the globalisation of labour market processes was beginning to develop from within the industrial complex. Firms specialising in the innovation and R&D stage of production, including start-up computer firms, continued to depend on localised core labour market processes, and were concerned about high labour turnover and a growing inability to fill skilled posts. The initial response to these problems of labour turnover and external recruitment was to raise wages, salaries and benefits. But this strategy led to wage inflation, encouraging the exodus of firms to low wage locations. Spatial restructuring also became problematic for the assembly operations of larger established firms, especially those involved in the production of proto-typical computer products. In response to global competition, it became increasingly important for these firms to integrate innovation and production more closely. This meant that, in order to survive, restructuring within the US semiconductor industry had to be driven by innovation rather than the search for cheaper labour (Angel, 1994).

Local firms have responded to these pressures by changing their internal management structures and intensifying external labor recruitment practices. The former strategy has involved a process of decentralisation or 'opening up' of internal labour markets in a way that stimulates entrepreneurship and innovation and promotes the free exchange of ideas within the complex in a climate of intense inter-firm competition (Saxenian, 1996). The latter strategy has been enabled by the influx of immigrant workers, who have offered a source of relatively cheap labour within the industrial complex. Firms have taken advantage

269

of new ethnic and gender divisions in the labour force. As a result, not only are predominantly women (64.4 per cent in 1986) employed in routine production work, but increasingly these are Asian (38.3 per cent) and Hispanic (17.0 per cent) rather than White (37.4 per cent) women.

As a result of these changing labour recruitment practices, Santa Clara County's labour-market processes have become increasingly segmented, a development which in fact was already evident in the early 1980s:

> Semiconductor production generated a bifurcated class structure in the county, one which was distinguished by a large proportion of highly skilled engineers and managerial personnel alongside an even larger number of minimally skilled manufacturing and assembly workers. A highly segregated residential pattern evolved to accommodate the vastly different nature of social reproduction required for these two dominant classes of labor power. As the industry expanded, it became increasingly difficult to accommodate and reproduce both segments of this dichotomized work force within the same metropolitan region. (Saxenian, 1984, p. 191)

Emerging contradictions, such as house price inflation, the growth of a local homeless population, a spatial imbalance between jobs and housing, environmental decay, and anti-growth sentiments among the elite work force, encouraged a collective (sectoral) response. Local executives organised the Santa Clara Manufacturers Association which recommended a range of local policies to address housing shortages, commuting problems and traffic congestion, land use planning conflicts, and the declining quality of life. As a result of these policy recommendations, Santa Clara County became one of the first local governments in the United States to enact controls on industrial development, with a view to enhancing the quality of local life and work (Saxenian, 1984).

The sectorally-led approach of the Santa Clara Manufacturers Association was further supported by the investment decisions of local firms specialising in a wide range of computer products. These firms have renewed their commitment to the region, transforming Silicon Valley into a elite control, research and proto-type production centre for their worldwide operations. Central to this strategy has been the adoption of neo-paternalistic labour practices, including investment in campus-like corporate headquarters facilities, and attempts to rebuild a sense of community and belonging among the work force despite increasing social polarisation and conflict within the sphere of labour reproduction.

In summary, the continuing success of the Silicon Valley industrial complex can be attributed to the operation of a dual set of labour control practices which have fostered labour market processes conducive to inter-firm mobility and innovation within the industrial complex, while at the same time maintaining the overall social and territorial coherence of the region despite the deepening of patterns of labour segmentation and resistance within it. These practices operate at

270

two levels in accordance with Silicon Valley's 'bifurcated class structure': at the level of core skilled workers dealing with the innovation side of production; and at the level of semi-skilled and unskilled assembly workers. In the case of core workers, there is a tradeoff between maintaining a rate of labour turnover within the industrial complex conducive to innovation and ensuring that highly qualified personnel are not recruited into rival centres outside the region. In the case of peripheral workers, the tradeoff takes place between recruiting from an increasingly segmented labour market (by ethnicity and gender) and ensuring that costs of labour reproduction do not escalate at a rate in disproportion to the region's capacity to reproduce labour power in that category. These tradeoffs have ensured that, in terms of the dominant labourmarket processes operating in the region, localisation tendencies continue to counteract globalisation tendencies, thereby reproducing in the medium term the territorial integrity and innovative capacity of the Silicon Valley production complex.

Control for exportation: maquiladoras and the devaluation of the peso

Nevertheless the endurance of areas like Silicon Valley as loci of production and innovation in the already industrialised countries is contingent upon labour market processes and labour control practices operating in competing localities in the industrialising countries, particularly those which are on the receiving end of foreign direct investment. One such area is the industrial border zone between the United States and Mexico, which was established to attract firms and industries away from US States like California. Despite their strategic involvement in the creation of the New International Division of Labour, even companies operating in the Mexican border zone are subject to localisation tendencies resulting from the imperatives of labour control.

Since its inception in 1965, Mexico's Border Industrialization Program has provided incentives for foreign (mainly US) manufacturers to locate labour intensive assembly operations in cities immediately south of the US border. By 1965 the border zone had a surplus of cheap labour, due in part to the ending of the Mexican Labor (Bracero) Program with the US and the tightening of US immigration laws. The devaluation of the peso in 1982 provided a further impetus for US manufacturers to establish assembly operations across the border. Mexican wages were less costly for US companies. In 1987, at $US 0.84 per hour, maquila wages were half what they had been in 1981. By 1988, over 1400 assembly manufacturers operated under the maquila programme, taking advantage of cheap labour costs and tariff policies to export assembled products to the US market (South, 1990). As a result of the 1995 peso devaluation, Mexico has become one of the cheapest locations in the world for labour intensive, export orientated assembly operations .

271

In terms of labour market processes, the maquila programme has been constructed around, first, US-Mexico wage differences and, second, labour control practices which minimise labour turnover within the maquila assembly plants. As products reach a mature stage in the life cycle, many US companies have chosen to locate final assembly operations in the border zone to take advantage of enclaves of non-union, cheap, and mainly female labour. The maquila programme has thus produced distinctive labour enclaves within the border cities which are characterised by low pay and exploitative working conditions. These employment practices are supported by national government policies which restrict wage increases and limit the bargaining powers of unions. Currently, Mexico's wage legislation places a 10 per cent limit on negotiated pay increases.

Despite these restrictions, many maquila employers offer fringe benefits to their workers in order to increase productivity and reduce labour turnover. Research has revealed that fringe benefits and payments-in-kind are generally more liberal in tighter local labour markets within the border manufacturing zone, particularly in cities having high concentrations of maquiladoras (and hence where unionisation is a potential threat) (South, 1990). The fact that assembly operations have tended to locate in the higher wage rate locations *within* the Mexican border zone further suggests that maquila employers not only recognise the importance of proximity to the US market and parent operations but also want to ensure there are adequate reserves of experienced, albeit cheap, labour for future expansion.

The presence of incipient labour market relations of reciprocity and flexible work practices has led to suggestions that the border zone offers some potential for growth along the lines of a flexible production system (Wilson, 1992). Indeed, when the zone was first established there was some expectation that foreign inward investment would foster local spin-offs and create a pool of capital and local entrepreneurial talent, and a level of technological development conducive to further growth. Nevertheless maquiladoras have tended not to develop strong local (forward and backward) linkages with other firms and sectors within the border zone, and the absence of local spin-offs (other than local services) has been a constraint on *in situ* industrialisation. More crucially, perhaps, the local labour control regime is geared primarily towards *ex situ* industrialisation. Many of the firms attracted to the border zone are labour intensive assembly operations which continue to use mass production techniques and despotic employment practices akin to 'bloody Taylorism'. In short, the labour control regime which currently prevails is more appropriate for 'global Fordism' than for a new regime of flexible accumulation.

This is not to argue that globalisation tendencies dominate local labour-market processes to the exclusion of localisation tendencies. The devaluation of the peso in 1995 revealed that there are limits to the extent to which despotic labour control practices can prevail, even in a locality clearly orientated towards production and exportation in the New International Division of Labour. The recent devaluation drove down maquila hourly wage rates further than previous devaluations, leaving many maquila employees struggling to obtain their basic daily needs of food, shelter and clothing. Since this clearly represented a threat to labour productivity, employees decided to intervene by supplementing their existing fringe benefit programmes. Workers were offered additional cash benefits, food coupons, and payments-in-kind to compensate for the loss of income. This strategy was essential for reproducing local labour power and minimising social unrest. At this basic level, then, the fostering of local labour market reciprocities has become a central aspect of the labour control regime which prevails along the US Mexico border.

Conclusion: new local-global geographies of labour resistance?

I have argued in this chapter that, far from there being historical tendencies towards the globalisation of labour control, the current period may be marked by the dominance of social and geographical counter-tendencies. In the wider context of economic globalisation, firms and industries have become increasingly dependent upon reciprocal labour market processes constructed around and within particular localities. At the same time, there is a great deal of inter-locality variation in the ways in which labour control practices operate. This variation reflects, on the one hand, the different functional roles (production strategies) of localities in the global economy and, on the other hand, how local actors (firms, industrial organisations, workers, public officials, residents, community organisations, etc) have responded to threats to those roles. To the extent that local responses have brought about the increasing localisation of capital (and, correspondingly, the localisation of labour), the tendency for firms and industries to restructure through space in response to labour control difficulties is prevented. This does not mean, however, that local labour control regimes are disconnected from wider systems or modes of social regulation. On the contrary, it is a necessary feature of uneven development that local variations in labour control regimes continue to persist in spite of the globalisation of control and regulatory frameworks in society at large.

The emphasis in this chapter has been on the relationship between labour control and firm and industry level production strategies. What, however, are the implications of the argument for understanding the changing social basis of resistance to the labour control practices of firms and industries? As Maguire (1988, p. 84) has argued, 'To concentrate excessively on (labour process change) misses the variety of managerial control strategies and the various ways workers comply with and resist these methods of control'. Clearly, geography, and in particular local variations in labour control practices, are central. I conclude therefore by highlighting four emerging geographical terrains of labour resistance.

Firstly, the constraints of localisation facing even globally competitive firms and industries suggest that it is too premature to conclude that place-based labour resistance has declined in importance in relation to the globalisation of production (Fainstein, 1987). The evidence that place or community-based movements continue to shape the landscape of production and control is simply too overwhelming to justify such a conclusion (Fitzgerald, 1991; Jonas, 1995). At the very least, it should be recognised that the nature of labour and community resistance to capital mobility is contingent upon the degree to which key players, including locally-based corporations, are locally dependent (Herod, 1991). And the level of resistance depends upon what kind of a threat is implied by globalisation and what sectors and interests in the local economy are most affected.

Secondly, the nature of localisation imperatives in flexible production centres means that qualitatively different labour and community networks from those that characterised mass production will be implicated in any resistance movement developing in those places.[8] While clearly many of these networks will be highly localised and industry-specific, others will transcend the boundaries of firms, industries and localities, locking work forces and community groups into increasingly globalised production networks. To what extent such local-global networks afford new opportunities for mobilisation around and against flexible production strategies remains an open question.

Thirdly, the emergence of new social divisions of labour and the deepening of patterns of segmentation by gender, race, ethnicity and skill within metropolitan labour markets is likely to throw up new axes of resistance outside the workplace. On the one hand, it should be possible to evaluate the extent to which workers are able to set aside the traditional tendency to separate work and community, and mobilise around consumption *and* production issues (Fitzgerald and Simmons, 1991). On the other hand, the extent to which the creation of new labour enclaves within metropolitan areas reinforces this tendency and further divides the work force is another possibility.

Fourth and finally, the extent to which agglomeration tendencies in flexible production systems create contradictions which increase the capacity for labour resistance should receive more attention. Flexible labour markets tend to be characterised by instability, high turnover, a lack of socialisation of skills, skills poaching and a variety of other internal contradictions (Peck, 1992). There are signs of growing resistance to flexible work practices within such production complexes,[9] suggesting the parameters of an emerging flexible system of labour regulation are in fact not yet in place, and therefore are likely to be shaped by ongoing local struggles and contexts. That such frameworks will develop *in situ* rather than 'from above' seems increasingly likely:

> Regulatory systems are not portable structures, achieving similar results wherever they are deployed, but are in fact deeply rooted in local social structures...Thus, it can be concluded that one of the root causes of the

variable nature of labour-market flexibility follows from the fact that these labour-market structures operate within a variety of national, regional, and local regulatory milieux...In short, the social and spatial context will have a major influence on the 'type' of labour-market flexibility which emerges in a region. (Peck, 1994a, p. 169)

In this regard, it should be noted that the relationship between flexibility and the changing geography of unionisation remains poorly understood. While there has been a decline in levels of unionisation in traditional (mass production) sectors like auto production, engineering, steel and shipbuilding, these have been partially offset by growth in other sectors, particularly service industries and the public sector (Allen, 1988). Since patterns of unionisation and de-affiliation vary sectorally and spatially, the geography of labour resistance to flexible work practices is likely to be uneven, reflecting the development of unique local cultures of unionisation built around 'uneven reserves' of labour (Wills, 1996). Far from contributing to the disorganisation and fragmentation of labour, the search for flexibility may create opportunities for labour to become more organised, especially in sectors and localities hitherto not noted for their high rates of unionisation or levels of resistance.

Of course it is important not to exaggerate these possibilities. We are a very long way from the point where a new national or even international labour movement can be reconstructed from the fragments of 'disorganised' labour. Resistance remains for the most community-bound and locality-specific as a result of, on the one hand, the historical legacy of local labour control practices and, on the other, the contemporary realities of 'hegemonic despotism'. Nor is it the case that locally-based resistance will necessarily engender more progressive and less despotic social contracts between labour and capital. Nevertheless, it is important at a theoretical level to recognise that localities continue to offer strategic sites for struggles around the 'social totality' of relations of production and reproduction.

Acknowledgements

This is a considerably revised version of a paper presented at the IGU Commission, Organisation of Industrial Space, Conference on "Interdependent and Uneven Development," Seoul, South Korea, August 1995. Thanks to National Science Foundation and the University of California for travel support, to the conference organisers for inviting me to participate, and to Ed Malecki for sound effects. Jamie Peck, Graham Haughton, Susan Hanson, Joan Fitzgerald and Suzy Reimer have in their own respective ways encouraged me to develop my ideas about labour control, but I alone am responsible for the contents of this chapter.

Notes

1.   See Peck (1994a; 1996) for a more general discussion of socially and spatially embedded systems of labour regulation.

2.   See Florida and Jonas (1991) for a general discussion of New Deal policy experiments.

3.   It is interesting to note that resurgent liberals have emphasized the centrality of reciprocal employment relations in rebuilding local and national economies. For example, Reich (1987, p. 248) has talked about the importance of 'reciprocal dependencies' between employers and workers in the context of US employment relations. The British Labour Party's notion of a 'stakeholder economy' is also suggestive.

4.   Other relevant concepts include 'regional structured coherence' (Harvey, 1985), 'factory regime' (Burawoy, 1975), 'local mode of social regulation' (Peck and Tickell, 1992), and 'local dependence' (Cox and Mair, 1988). For a critical discussion of the relationships between these concepts and 'local labour control regime' see Jonas (1996).

5.   For example, institutions involved in the after-Fordist regulation of labour markets and employment relations might be more appropriately researched at the regional scale (Perulli, 1993).

6.   Compare with the use of typologies in the 'new urban politics' literature (Cox, 1993).

7.   John Schafer, personal interview, 1993. Since leaving Norton Company to work for another Worcester-based company, however, Mr Schafer professes to be an advocate of unions.

8.   Under mass production, there was a tendency to separate at an ideological level the business enterprise from the community. Community politics and social networks were conceived of as 'things' to 'become involved in' rather than as comprising integral elements of a local production strategy. Not surprisingly the advent of the large vertically integrated enterprise was often viewed as a threat to traditional community networks and social relations (Gutman, 1976).

9.   Pollard (1995) has examined local responses to labour flexibility in the Los Angeles retail banking industry. Since 1987, a number of large retail banks in Los Angeles have replaced full time clerical positions with part time and temporary positions. While this has helped to lower labour

costs for the banks, it has also created problems of labour control and high rates of labour turnover. Although retail banking has traditionally been a non-union industry in which predominantly female clerical labour is employed (see also Wills, 1996), there is some evidence of organised resistance to the adoption of flexible work practices by the industry.

References

Aglietta, M. (1979), *A Theory of Capitalist Regulation*, New Left Books, London..

Allen, J. (1988), 'Fragmented firms, disorganized labour?', in Allen, J. and Massey, D. (eds) *The Economy in Question*, Sage, London, pp. 184-267.

Angel, D.P. (1994), *Restructuring for Innovation: The Remaking of the U.S. Semiconductor Industry*, The Guilford Press, London.

Appelbaum, E. and Batt, R. (1994), *The New American Workplace,* ILR Press, New York.

Berberoglu, B. (ed.) (1993), *The Labor Process and Control of Labor: The Changing Nature of Work Relations in the Late 20th Century*, Praeger, Westport, CT.

Best, M. (1990), *The New Competition,* Harvard University Press, Cambridge, MA.

Bluestone, B. and Harrison, B. (1982), *The Deindustrialization of America: Plant Closings, Community Abandonment, and the Dismantling of Basic Industry*, Basic Books, New York.

Burawoy, M. (1979), *Manufacturing Consent*, University of Chicago Press, Chicago.

Burawoy, M. (1985), *The Politics of Production: Factory Regimes Under Capitalism and Socialism*, Verso, London.

Cheape, C. (1985), *Family Firm to Modern Multinational: Norton Company, a New England Enterprise*, Harvard Studies in Business History, Cambridge, MA.

Cohen, B. (1988), 'The Worcester machinists' strike of 1915', *Historical Journal of Massachusetts*, vol. 10, pp. 154-71.

Cox, K.R. (1993), 'The local and the global in the new urban politics: A critical view', *Environment and Planning D: Society and Space*, vol. 11, pp. 433-448.

Davis, M. (1986), *Prisoners of the American Dream,* Verso, London.

Dicken, P. (1976), 'The multiplant enterprise and geographical space: some issues in the study of external control and regional development', *Regional Studies*, vol. 10, pp. 401-12.

Dicken, P. (1992), *Global Shift: The Internationalization of Economic Activity*, The Guilford Press, New York.

Edwards, R. (1979), *Contested Terrain: The Transformation of the Workplace in the Twentieth Century*, Basic Books, New York.

Fainstein, S.S. (1987), 'Local mobilization and economic discontent', in Smith, M.P. and Faegin, J.R. (eds) *The Capitalist City: Global Restructuring and Community Politics*, Blackwell, Oxford, pp. 323-42.

Fitzgerald, J. (1991), 'Class as community: the new dynamics of social change', *Environment and Planning D: Society and Space*, vol. 9, pp. 117-28.

Fitzgerald, J. and Simmons, L. (1991), 'From consumption to production: Labor participation in grassroots movements in Pittsburgh and Hartford', *Urban Affairs Quarterly,* vol. 26, pp. 412-531.

Florida, R.L. and Jonas, A.E.G. (1991), 'U.S. urban policy: The postwar state and capitalist regulation', *Antipode,* vol. 23, pp. 349-384.

Gartman, D. (1993), 'The historical roots of the division of labor in the U.S. auto industry', in B. Berberoglu (ed.) *The Labor Process and Control of Labor : The Changing Nature of Work Relations in the Late 20th Century,* Praeger, Westport, CT, pp. 21-43.

Gordon, D.M. (1984), 'Capitalist development and the history of American cities', in Tabb, W.K. and Sawers, L. (eds) *Marxism and the Metropolis,* Oxford University Press, New York, pp. 21-53.

Gordon, D.M., Edwards, R. and Reich M. (1982), *Segmented Work: Divided Workers: The Historical Transformation of Labor in the United States,* Cambridge University Press, Cambridge.

Granovetter, M. and Tilly, C. (1988), 'Inequality and labor processes' in Smelser, N.J. (ed.), *Handbook of Sociology,* Sage, Beverley Hills, pp. 175-221.

Gutman, H. (1976), *Work, Culture and Society in Industrializing America,* Alfred A. Knopf, New York.

Hanson, S. and Pratt, G. (1992), 'Dynamic dependencies: a geographic investigation of local labour markets', *Economic Geography,* vol. 68, no. 4, pp. 373-405.

Hanson, S. and Pratt, G. (1995), *Gender, Work and Space,* Routledge, London.

Hareven, T. (1982), *Family Time and Industrial Time: The Relationship Between the Family and Work in a New England Industrial Community,* Cambridge University Press, Cambridge.

Harvey, D.W. (1985), *The Urbanization of Capital,* The Johns Hopkins University Press, Baltimore, MD.

Harvey, D.W. (1989), *The Condition of Postmodernity,* Blackwell, Oxford.

Herod, A. (1991), 'Local political practice in response to a manufacturing plant closure: how geography complicates class analysis', *Antipode,* vol. 23, pp. 385-402.

Herod, A. (1994), 'Further reflections on organized labor and deindustrialization in the United States', *Antipode,* vol. 26, pp. 77-95.

Jessop, B. (1990), *State Theory: Putting the Capitalist State in its Place,* Polity Press, Cambridge.

Jessop, B. (1994), 'The transition to post-Fordism and the Schumpeterian workfare state', in Burrows, R. and Loader, B. (eds) *Towards a Post-Fordist Welfare State?* Routledge, London, pp. 13-37.

Jonas, A.E.G. (1992), 'Corporate takeover and community politics: the case of Norton Company in Worcester', *Economic Geography,* vol. 68, pp. 348-372.

Jonas, A.E.G. (1995), 'Labor and community in the deindustrialization of urban America', *Journal of Urban Affairs,* vol. 17, pp. 183-199.

279

Jonas, A.E.G. (1996), 'Local labour control regimes: Uneven development and the social regulation of production', *Regional Studies*, vol. 30, pp. 323-338.

Kowal, D. (1990), 'BASF uncovers its wraps', *Worcester Magazine*, April 18, pp. 6-7.

Lipietz, A. (1987), *Mirages and Miracles: The Crises of Global Fordism,* Verso, London.

Littler, C. (1982), *The Development of the Labour Process in Capitalist Societies*, Heinemann, London.

Lockwood, D. (1966), 'Sources of variation in working class images of society', *Sociological Review*, vol. 14, pp. 249-267.

Mair, A., Florida, R. and Kenney, M. (1988), 'The new geography of automobile production: Japanese transplants in North America', *Economic Geography*, vol. 64, pp. 352-73.

Maguire, M. (1988), 'Work, locality and social control', *Work, Employment and Society*, vol. 2, no. 1, pp. 71-87.

Manwaring, T. (1984), 'The extended internal labor market', *Cambridge Journal of Economics*, vol. 8, pp. 161-87.

Massey, D. (1984), *Spatial Divisions of Labor: Social Structures and the Geography of Production,* Methuen, New York.

Melling, J. (1992), 'Employers, workplace culture and workers' politics: British industry and workers' welfare programmes', in Melling, J. and Barry, J, (eds) *Culture in History: Production, Consumption and Values in Historical Perspective,* University of Exeter Press, Exeter, pp. 109-36.

Nash, J. (1987), 'Community and corporations in the restructuring of industry', in Smith, M.P. and Faegin, J.R. (eds) *The Capitalist City: Global Restructuring and Community Politics*, Blackwell, Oxford, pp. 275-296.

Newby, H. (1977), *The Deferential Worker,* Allen Lane, London.

Norris, G.M. (1978), 'Industrial paternalist capitalism and local labour markets', *Sociology*, vol. 12, pp. 469-490.

Peck, J. A. (1992), 'Labor and agglomeration: Control and flexibility in local labor markets', *Economic Geography*, vol. 68, no. 4, pp. 325-347.

Peck, J. A. (1994a), 'Regulating labour: The social regulation and reproduction of local labour-markets', in Amin, A. and Thrift, N. (eds) *Globalization, Institutions and Regional Development in Europe*, Oxford University Press, Oxford, pp. 147-176.

Peck, J.A. (1994b), 'Localising labour: neoliberalism and the decentralisation of labour regulation', paper presented at the 90th annual conference of the Association of American Geographers, San Francisco, March.

Peck, J.A. (1996), *Workplace,* The Guilford Press, New York.

Peck, J.A. and Tickell, A. (1992), 'Local modes of social regulation? Regulation theory, Thatcherism and uneven development', *Geoforum*, vol. 23, pp. 347-363.

Perulli, (1993) 'Towards a regionalization of industrial relations', *International Journal of Urban and Regional Research*, vol. 17, pp. 98-113.

Piore, M. and Sabel, C. (1984), *The Second Industrial Divide*, Basic Books, New York.

Pollard, J.S. (1995), 'The contradictions of flexibility: labour control and resistance in the Los Angeles banking industry', *Geoforum*, vol. 26, pp. 121-138.

Reich, R. (1987), *Tales of a New America*, Vintage, New York..

Rosenzweig, R. (1983), *Eight Hours for What We Will: Workers and Leisure in an Industrial City, 1870-1920*, Cambridge University Press, New York.

Saxenian, A. (1984), 'The urban contradictions of Silicon Valley: regional growth and the restructuring of the semiconductor industry', in Sawers, L. and Tabb, W.K. (eds) *Sunbelt/Snowbelt: Urban Development and Regional Restructuring*, Oxford University Press, New York, pp. 163-197.

Saxenian, A. (1994), *Regional Advantage: Culture and Competition in Silicon Valley and Route 128*, Harvard University Press, Cambridge, MA.

Saxenian, A. (1996), 'Inside-out: regional networks and industrial adaption in Silicon Valley and Route 128', *Cityscape*, vol. 2, no. 2, pp. 41-60.

Scott, A.J. (1988) *New Industrial Spaces: Flexible Production Organization and Regional Development in North America and Western Europe*, Pion, London.

Scott, A.J. and Angel, D.P. (1988), 'The global assembly operations of U.S. semiconductor firms: a geographical analysis', *Environment and Planning A*, vol. 19, pp. 875-912.

South, R.B. (1990), 'Transnational "maquiladora" location', *Annals of the Association of American Geographers*, vol. 80, pp. 549-570.

Storper, M. and Scott, A.J. (1989), 'The geographical foundations and social regulation of flexib production complexes', in Wolch, J. and Dear, M. (eds) *The Power of Geography: Territory Shapes Social Life*, Unwin Hyman, Boston, pp. 21-40.

Storper, M. and Walker, R. (1983), 'The theory of labor and the theory of location', *International Journal of Urban and Regional Research*, vol. 7, pp. 1-43.

Storper, M. and R. Walker, (1989), *The Capitalist Imperative: Territory, Technology and Industrial Growth*, Blackwell, New York.

Tickell, A. and Peck, J.A. (1995), 'Social regulation after Fordism: Regulation theory, neo-liberalism and the global-local nexus', *Economy and Society*, vol. 24, pp. 357-86.

Warde, A. (1988) 'Industrial restructuring, local politics and the reproduction of labour power: Some theoretical considerations', *Environment and Planning D: Society and Space*, vol. 6, pp. 75-95.

Warde, A. (1989), 'Industrial discipline: Factory regime and politics in Lancaster', *Work, Employment and Society*, vol. 3, pp. 49-63.

Wills, J. (1996), 'Uneven reserves: Geographies of banking trade unionism', *Regional Studies*, vol. 30, pp. 350-362.

Wilson, P.A. (1992), *Exports and Local Development: Mexico's New Maquiladoras*, University of Texas Press, Austin.

Zunz, O. (1982), *The Changing Face of Inequality. Urbanization, Industrial Development, and Immigrants in Detroit, 1880-1920,* University of Chicago Press, Chicago.

12 So what is internationalisation? Lessons from restructuring at Australia's 'mother plant'

Phillip M. O'Neill

Introduction

This chapter examines the decline of the Newcastle steelworks in order to investigate the meanings of *internationalisation* that are attached to the restructuring process. Initially, the range of meanings of internationalisation are explored in general. The chapter then proceeds through a detailed examination of the processes of restructuring at the Newcastle steelworks in the context of a decade of restructuring of its operator, Broken Hill Proprietary Company Limited (BHP), Australia's largest and most internationalised corporation. In 1994, it ranked 126 on the Fortune Global 500 list of the world's largest industrial companies. BHP is divided into three main international business groups - BHP Minerals, BHP Petroleum and BHP Steel. In 1995, the corporation managed an annual cash flow of over US$12 billion, a third of which was generated in Australia. It has assets of about US$20 billion with operations in more than 20 nations.

The company commenced operations as a silver, lead and zinc miner at Broken Hill in the desert regions of western New South Wales. There it began a tradition of militant labour relations, strong (though not necessarily compassionate) community ties, and intimate collaborations with state apparatus. With the decline of the Broken Hill ore deposit, BHP started its reign as Australia's monopoly steelmaker in 1915 by opening an integrated steelworks at Newcastle, 150 kilometres north of Sydney. The works was often referred to as BHP's 'mother plant': it was a constant and reliable supplier of feedstock for an Australia-wide network of rolling mills and it performed the role of nursery for the emergence of future technicians and managers. The image of the mother plant was also a metaphor for twentieth century industrial growth in Australia: strong, large scale development based on processing native raw materials. Steelmaking became BHP's core business and BHP became Australia's industrial standard bearer. By the 1960s, the company had added large integrated steelworks at Port Kembla, south of Sydney, and at Whyalla in South Australia. These were integrated vertically with downstream processors, wholesalers and distributors under the BHP banner. All that could be produced, and more, was sold into the world's most protected post war economy. Capital accumulation was spectacular in size, enduring and guaranteed.

In the early 1980s, however, the profitability of Australian steelmaking declined dramatically. Production technologies had become outdated; returns to organised labour were above world tolerances; downstream demand in the protected Australian economy declined markedly; cheaper imported steel from East Asia was readily available; and community intolerance of local air pollution was mounting. BHP's response was dramatic. A major restructuring of its Australian steel operations began. With assistance from a federal Steel Plan, BHP re-invested heavily in its Port Kembla works installing new steelmaking technologies and capacities to service new, tied mills in south-east Asia and along the US west coast. Simultaneously, major reductions occurred in employment levels at the three Australian steelworks. The Newcastle steelworks suffered a major fall in status being reduced to the supply of vanilla-grade steel products to a stagnant domestic market. Where once over 11,000 workers produced steel for over 10,000 others in foundries, mills and engineering works, these 21,000 jobs are done by just 8,000 workers - 3,000 of whom remain in the old integrated works. The works has been under threat of closure since 1982. Stories of shut down have raced through the Newcastle community with increasing frequency.

During the same period, the works' management has assembled a number of bids to BHP's corporate headquarters for investment finance - involving new stories of hope for projects to propel the works to a new, international role. Yet only minimal maintenance funding has been applied. Besides demand for the financing of investments at the Port Kembla steelworks, competition for funds has also come from BHP's petroleum and minerals divisions which have become pathways for the corporation's major investments offshore. This thrust commenced in 1984 through the acquisition of international minerals and petroleum producer Utah International Inc (USA) which secured the enormous Escondida copper works in Chile - far and away the most important contributor to BHP profitability in the mid 1990s. Other important acquisitions have included the North American petroleum producer Energy Reserves Group Inc in 1985 and the European-based Hamilton Oil Corporation in 1991.

Meanings attached to internationalisation

In orthodox political economy, internationalisation is usually seen as a key strategy for monopoly capitalists to resolve the problem of over-production and domestic market saturation (Baran and Sweezy 1966, Mandel 1978, Harvey 1982). In this view, internationalisation is advanced as an economic process

284

which relies on a growing mobility of capital to temporarily resolve the illogicality of capitalism. Consequently, it presents investments in old industrial regions as increasingly vulnerable to closure with local labour forced to bargain away the wages and conditions attained in previous growth decades (for example Hudson and Sadler, 1986). In a variation, Dicken et al (1997) argue that the term internationalisation should be reserved for specified processes of expansion by enterprises into foreign markets and production spaces. They contrast the term *globalisation* which they claim is discursively constructed to serve the objectives of special interest groups especially those seeking to undermine previously negotiated conditions of work. The problem with this neat semantic division, however, is that it cannot be imposed on all those individuals and groups who are engaged in argument about the direction of economic change. This chapter demonstrates that, like globalisation, internationalisation holds different meanings for different actors. In many instances it is used interchangeably with globalisation. Further, it cannot have a technical or academic meaning separate from the coercive effects of the wider discourses in which the term is employed.

So internationalisation is analysed here not as a explicit economic event or process but rather as an inseparable part of the discourse of economic change. For its users, it represents an argument about the direction of economic change. It evokes scenes involving interplay among forces from many different scales involving a variety of agents. The construction of these images of contest is used to coerce groups into the support and adoption of specific courses of action. In the case of BHP, four different meanings of internationalisation may be associated with the company's restructuring of its steel division (Fagan and O'Neill, forthcoming). Identifying these four meanings not only assists understanding the motivations and directions of ensuing restructuring processes but may also aid the mobilisation of better targeted intervention and resistance tactics.

The first meaning of internationalisation identified is based on inductive logic of the market place and the need for growth. It is explained by Dicken's arguments (1992, 1994) that there are identifiable product and factor market advantages in organising production across national boundaries. It is an extension of the argument that internationalisation is, at least, a short-term solution to the inherent contradiction of monopoly capitalism. It is a meaning that is commonly found in academic texts, government policy documents and official corporate reports.

The second meaning presents internationalisation as an organisational process enabled by new transport and communications technologies and the international entanglement of management systems resulting from cross-border takeovers, strategic alliances, new supra-national regulatory continuities and national regulatory discontinuities (see Dicken et al, 1994). This meaning is predominant in normative models of business behaviour as well as in populist accounts of future directions of economic change.

A third meaning of internationalisation elevates the importance of the international financial transaction and the role of the corporate treasurer in managing a portfolio of investments ranging from long-term productive outlays to short-term harvestable securities. Stories of the internationalisation of corporate financial processes are told to demonstrate the power of the agents of capital to forego longer term productive investments should opportunities arise for speculative short-term gain. They are told in order to enhance the power of corporate financial managers who seek to centralise corporate decision making about the reinvestment of surpluses. In addition, the financially-based discourse of internationalisation imposes a discipline on the distributional decisions of production-based managers away from the corporate centre by providing strict financial guidelines for the containment of costs and the distribution of revenues.

A fourth meaning of internationalisation lies in its use as a general public discourse about restructuring. It has been long recognised by workers and communities in older industrial regions but less so in the literature (cf. Webber et al, 1991; Clark et al, 1992). Internationalisation, as a public discourse perpetrated by governments, corporations, labour unions and others, plays a major role as a coercive tool in generating compliance with restructuring ventures. Internationalisation as coercion is most evident at two scales here: its use by the Australian state to secure the compliance of nationally organised labour in the liberalisation of product and factor markets (O'Neill, forthcoming); and its use by large corporations, including BHP, to accelerate workplace reforms at specific production sites. In particular, this involves the intensification of work to maximise the returns from the operation of near-to-exhausted steel-making plant.

Internationalisation and restructuring

Each of these four meanings of internationalisation permeates the restructuring of BHP's steel division as the corporation has expanded its investments in newer, foreign-based minerals and petroleum divisions. The meanings provide each of context, need and opportunity for rationalisation of domestic steelmaking operations. The meanings have a major influence on the development of new productive strategies. They determine the timing of coercive moments. They set the financial yardsticks to be met by BHP's domestic producers (set obstinately at 15 per cent p.a. during the last decade). They dictate the organisational structures to meet the demands of production, marketing and management. And they help generate the public discourse which guides and shapes the responses of individuals and groups.

These meanings of internationalisation are applied here to the historical contests and tensions involved in the structuring and restructuring of production at the Newcastle steelworks. A counter argument is developed to the view that internationalisation is a general economic process and one that overrides processes at other scales. *First*, it is shown how accumulation crises derive from the physically embedded nature of local events and agencies in the processes of capital accumulation. *Second*, the strategy of managing the extraction of value from an ageing plant is shown to derive from the role played by *residual value* not just sunk costs or off-shore competition. *Third*, close links are identified between restructuring strategies and domestic market structures. *Fourth*, just as the organisation of new investments involves careful maintenance of state-capital relations, so it is found that political relationships remain critical to the ongoing operation of ageing production facilities. *Fifth*, the manipulation of the public discussion of restructuring is shown to play a critical role in the management of operations during a residual value period. *Sixth*, it is demonstrated that, in producing and reacting to the meanings of internationalisation, managers privilege one or more of the multiple class positions they hold which in turn has a major bearing on the directions of restructuring, on plant operations and the capacity of local labour to have significant influence on distributional outcomes.

Accumulation, territorial linkages and crisis at the Newcastle steelworks

A crisis in accumulation cannot be reduced simply to the single-dimension problem of a monopoly capitalist outgrowing a domestic market, nor of local change being determined by national and global scale forces. Accumulation is an historical process involving past events which were global, national and local in scale. The contemporary local, then, embodies the outcomes of all past processes, whatever their original scale of formation. The establishment of the Newcastle steelworks involved the *structuring* of local linkages through events situated at national and international scales. The founding manager of the Newcastle works:

> ... stressed the importance of developing a large steel producing capacity to take advantage of economies of scale, but suggested that in order to avoid an unwieldy organisation [BHP] should encourage associated and subsidiary industries to establish themselves close to the works. Then, "the Company, being the controller of the steel supply, could purchase any of the finishing plants when a favourable opportunity offered." (Hughes 1964, p.81, quoting establishment engineer David Baker)

BHP general manager Guillaume Delprat affirmed that Baker's view was corporate policy in a public statement in 1920:

> We wish to establish other firms alongside us in order to secure a constant and regular outlet for our steel. We enter into contracts with firms possessed of experience in making special lines and guarantee their supply of raw material...We aspire to be the Mother, or Key industry of many dependent industries valuable and necessary to the growth and maturity of the Australian nation. (cited in Hughes, 1964, p. 82)

Upstream suppliers, service providers and downstream processors commenced operations in Newcastle in association with the creation of the steelworks. The Austral Nail Co Ltd of South Melbourne, for example, was approached to build a wire plant in Newcastle in 1917, and was followed soon after by UK companies Ryland Bros and Lysaght Bros which were Australia's main wire suppliers (Hughes 1964). Similarly, by being able to guarantee exclusive access to the growing Australian market, BHP attracted local steel fabrication investments by three English steel companies, Bullivants (wire rope), Vickers (wheels and axles) and Stewarts and Lloyds (pipes and tubes). During the following two decades, in its drive to establish a vertically integrated Australian steel industry under monopoly control, BHP acquired all the Newcastle companies with which it had major linkages. It consolidated ownership of its Hunter Valley coal suppliers by acquiring the John Darling colliery in 1925 and the Lambton and Burwood collieries in 1932. BHP entered the shipping industry in 1925 ending its reliance on British firms and averting rising shipping costs and a growing incidence of industrial disputes in maritime industries - BHP was a steel producer whose activities relied on unbroken supply and continuous production. The Australian subsidiaries of Bullivants, Vickers, and Stewarts and Lloyds were acquired by BHP between 1929 and 1935. In 1932 it acquired its rival Port Kembla raw steel producer, Australian Iron and Steel.

The Newcastle steel ensemble remained intact for the next half century. It thrived in the decades following the Second World War: imports of manufactured goods with high steel content were effectively prohibited by 'stringent' trade controls instituted by the Menzies Liberal government (Hughes, 1964) and by the protection of downstream Australian steel users (Lewis and McDonald, 1986). By the end of the 1970s, BHP had positioned itself firmly in the centre of the production and consumption of Australian steel. It supplied around ninety per cent of the domestic steel market. Forty per cent of its domestic sales was shipped to BHP subsidiary companies and another twenty per cent was taken by companies in which it had financial interests or contractual purchase arrangements. Steel exports were *opportunistic* rather than *strategic*: BHP was a domestic steel company whose operations were underpinned by an immense political power which it used cleverly to secure privileged access to high-quality natural resources (Fagan, 1984, 1986, 1988; Donaldson, 1981; Donaldson and Donaldson, 1983). It was a major employer and raw-steel producer in both Newcastle and Port Kembla in New South Wales and Whyalla in South Australia. It operated its own coal and iron ore mining and processing facilities and a vast network of capital city steel-processing mills (Rich, 1984). Many Australian local economies were highly dependent on the company's steelmaking activities and Newcastle experienced prolonged regional prosperity. In a national economy where everything made (and more) could be sold, steel industry unions held a powerful bargaining position and achieved substantial improvements in real wages despite BHP's history of antagonistic opposition to organised labour.

An accumulation crisis descended on the Australian steel industry in the early 1980s. BHP attributed the crisis to factors exogenous to the company: world-wide steel surpluses, depressed prices, over-valued currency and domestic recession (BHP Steel 1989). Of course, BHP argued publicly that these factors rendered unsustainable the costly management and work practices which had evolved during the post war growth period (Lewis and McDonald,1986). Severe economic and social upheaval ensued in Australia's steel regions. A pig iron blast furnace at Kwinana (WA) was closed in 1982. Major cost reduction programmes accompanied production and plant rationalisation and the steel towns. Newcastle, Wollongong and Whyalla, suffered enormous social dislocation (Schultz, 1985). Between 1982 and 1985, 14,701 jobs were lost in BHP's steelworks and coking-coal collieries (Fagan, 1988). In Newcastle during this period, 4,943 steelworks jobs were shed. Surprisingly, BHP continued steelmaking in all three major centres - an unusual approach to restructuring in a multi-locational manufacturing firm (Stafford and Watts, 1991). Yet there are problems with the abandonment of steelmaking in a large plant. Bradbury (1987) notes that the closure of large integrated steelworks is contrary to the convention of maintaining a pattern of constant reproduction of part or whole of fixed steelmaking capital. Because steelmaking is capital intensive, and its machinery is durable and expensive to maintain, a steelworks requires continuous, long-term operation to yield acceptable returns.

There is a further reason why BHP retained steelmaking at the three centres. It relates to BHP's strategic use of political power. The onset of recession was incidental to BHP's intentions to commence major steel plant rationalisation and technological change. During the 1970s, capital investment at BHP's steelworks had fallen behind the rate achieved by leading overseas producers (Donaldson and Donaldson, 1983; Australia. Steel Industry Authority, 1988). In 1983, for example, BHP produced less than 30 per cent of its output by the continuous casting method compared to a world-wide industry standard of 70 per cent (Australia. Steel Industry Authority, 1988). Recession provided the opportunity for BHP to argue successfully to the newly elected Hawke-Labor Government for subsidies towards the cost of restructuring (Larcombe, 1983). BHP threatened publicly that it would close the Newcastle steelworks unless the steel industry received restructuring assistance in the form of increased protection and capital subsidy (Garlic and Skaines, 1983).

BHP's ability to carry out threats to close its Australian steelmaking operations was a crucial factor in securing state assistance and workforce compliance in its restructuring programmes. This capability derived from BHP's emergence during the 1970s as a multi-commodity, multi-locational firm. The introduction of the Steel Industry Plan in 1983 represented a generous subsidy to capital and a handsome victory for BHP's political strategists. Between 1984 and 1988, the Australian government directly assisted BHP Steel's capital renewal programme through increased domestic product market protection, accelerated tax depreciation arrangements, bounty payments to domestic steel users, and regional employment programmes to aid community acceptance of restructuring. The Steel Industry Plan was supervised by a tripartite group from BHP, government, and trade unions. It legitimised massive reductions in employment and new technical and replacement investments throughout BHP Steel's Australian operations.

Critically, it secured the compliance of the steel unions through a series of labour relations agreements commencing with the Steel Industry Plan Agreement and followed by the Steel Industry Development Program Agreement (SIDA) between 1989 and 1992, and then by site agreements under the National Steel Industry Business Improvement Agreement (NSIBIA). In agreeing to the Steel Industry Plan, BHP Steel undertook to continue steel production at the three integrated steel centres, and ensured that employment reductions would occur by natural attrition and voluntary retirement and not by retrenchment. In return, steel industry unions gave commitments to new, non-militant, dispute settling procedures. Importantly, the Steel Industry Plan and its successors enabled BHP Steel to successfully implement a massive cost reduction programme, the Business Improvement Program, (BIP) which overlayed and intensified other negotiated changes (see below).

Once state assistance was secured, BHP invested confidently in its steel division, allocating an initial four billion dollars for plant upgrade and expansion (*Australian Financial Review*, 25.6.92, p.14). In conjunction with considerably lower employment costs, BHP Steel returned quickly to profitability. Domestic plant once more operated at full capacity diluting the high fixed costs of integrated steelworks capital and securing BHP's monopoly market position. Utilisation of the Australian steelworks was further bolstered by the establishment of over forty offshore roll-forming and painting lines. In addition, market contestability was diminished by pricing strategies which repelled imports, the production of a full range of steel products and by the control of capital city steel distribution.

The restructuring of production and market relations was accompanied by corporate restructuring. In 1985, BHP Steel was divided into seven major businesses based on alignments between existing domestic product market segments and the company's international marketing and production ambitions (Figure 12.1). BHP publicly explained the restructuring as the steel division's participation in the corporation's internationalisation strategy - even though the division's continued profitability remained largely dependent on regulated Australian domestic demand in a market place characterised by an *absence* of international market contestability.

The restructuring produced a new spatial division of labour within BHP Steel. National product market coverage was achieved by dedicating particular product ranges to each of the integrated steelworks: rods and bars at Newcastle, flat and coiled steel at Port Kembla, and long structural steel and rails at Whyalla. The specialisations strengthened the correlation between geography and product and formed the basis of BHP Steel's business divisions, enabling the devolution of responsibility for rationalisation and intensification of production, and marketing, to meet the financial targets set by either the steel group or by the wider corporation.

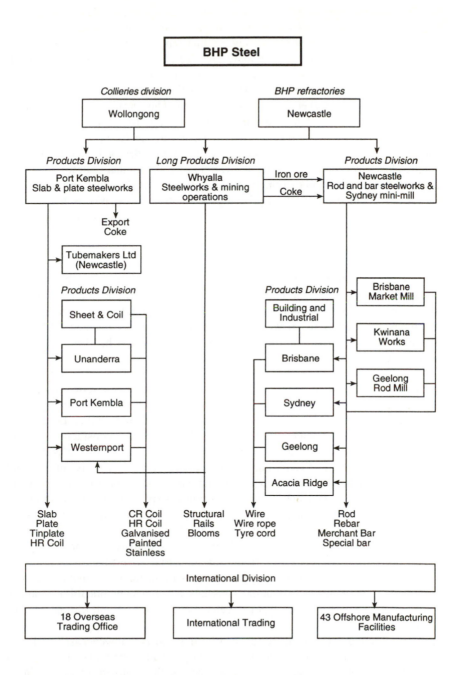

Figure 12.1 BHP Steel's interdivisional relationships

291

Thus output from the Rod and Bar Products Division (RBPD), based on the Newcastle steelworks, was confined to low value-added products for the construction industry, being the narrow product range around which Newcastle's steel producing ensemble was originally established (Fisher and Smith, 1980). It also closely matched the output range of electric arc furnace (EAF) mini mills and brought it into direct competition with an EAF mini mill in outer Melbourne which had been established in 1984 (see below). The world-wide expansion of mini mills also meant there was limited opportunity for *strategic* exports of surpluses. Thus, the accumulation crisis persisted at the Newcastle works. The conditions for successful accumulation over many decades had been created by complex historical forces operating at a variety of geographical scales. Crisis was, similarly, a complex interaction of technological, distributional, political and market-based processes. Not surprisingly, a solution to crisis was not readily forthcoming.

Harvesting 'residual value'

BHP Steel's reluctance to undertake major refit or replacement of Newcastle's steelmaking plant during the mid and late 1980s promoted uncertainty about the plant's future. This uncertainty has been harnessed by BHP to produce the conditions for harvesting *residual value* from the works through workplace intensification activities. The rhetoric of internationalisation has been central to the accompanying campaign of coercion.

The processes of capital exhaustion and firm exit from a large-scale manufacturing investment are described in many ways in the literature. Workplace intensification is commonly identified (Hudson, 1989; Scherrer, 1991) especially in conjunction with plant closures elsewhere (Stafford and Watts, 1991). The decision to completely abandon a market involves the wider consideration of an enterprise's financial structures and obligations (Clark, 1988). MacLachlan (1992), Clark (1992) and Clark and Wrigley (1993) have drawn insight into exit behaviour from the Schumpeterian formulations of the role of sunk costs in influencing firms' market behaviours (for original explanations see Schumpeter, 1947, especially p. 157 and Baumol, et al 1988). Clark and Wrigley (1993) propose the use of sunk costs as a vehicle for understanding the apparent divergence in the restructuring paths followed in old industrial regions compared to new industrial spaces. Firms with old industrial investment sites, having pursued internal scale economies during periods of sustained economic growth, are seen to have incurred substantial sunk costs within their portfolio of assets which severely constrain their market behaviours, including the ultimate act of market exit itself.

Consideration of the importance of sunk costs, however, should not overwhelm the analysis of what happens in a large plant when it becomes obvious (at least to its corporate parent) that it will not be the recipient of further large scale investment. Drawing from a rapidly expanding literature on the practice of

292

corporate treasury (for example Chew, 1993; Goold et al 1994; Love, 1995), it is argued here that in addition to consideration of exit costs, corporate treasuries are increasingly aware of the harvestable cash flows (or *residual values*) available from fully amortised (yet still operating) investments (Figure 12.2). At the same time, the operators of this plant (the branch or divisional managers) are precluded from accessing substantial new investment monies by a number of factors including: shortened forecasting periods (aligned with uncertainty of future yields in financial markets); shortened periods for the calculation of yields from productive investments (requiring higher rates to enable amortisation); denial of residual value beyond the forecast period (potentially a lucrative source of earnings); and over-valued exit costs (valued prematurely at the end of the forecast period, rather than at the end of the residual value period).

Not surprisingly, BHP's corporate treasury is keenly aware of the ongoing value that is able to be harvested from intensified work at the Newcastle steelworks. In this context, local management has been unsuccessful in seeking access to investment funding to enable transfer to a new investment cycle (see O'Neill and Gibson-Graham, 1997). Funds available to RBPD have been directed away from the steelworks into activities designed to protect domestic market share (see below). Domestic steel dispatches from the Newcastle steelworks have fallen consistently during the early 1990s with surpluses sold at discounted prices on world markets. It has been clear since the early 1990s that the Newcastle steelworks had no chance of attracting any new investment funding. Indeed, in the residual value period, even maintenance and upgrades within RBPD have had to be internally financed.

Local territorial linkages were dismantled as the steel ensemble lost purpose. Many BHP activities in Newcastle were abandoned in the early 1990s. The administration of BHP Collieries was transferred to Port Kembla and shipping fleet management was transferred to Melbourne; BHP's Hunter Valley coal mines and its high value-added steel processor, Commonwealth Steel were sold; the local skelp mill and foundry were closed and key local customers supplied with feedstock from the Port Kembla works; even local surplus land has been redeveloped for housing and coastal resort development.

Importantly, the closures and transfers did not result from a *planned* downgrading of Newcastle as a steel production centre but, rather, represented independent decisions of different operating divisions in the process of rebuilding and rationalising their own production systems.The losses were incremental as local BHP management struggled to meet the performance targets achieved more easily by other BHP operations. Hence, while the restructuring of national steel production within BHP involved the strengthening of vertical links between domestic raw materials and basic steel production to internationalised processing and marketing activities, opportunities for participation in the grand international plan by activities based in Newcastle were gradually eliminated. Yet, the

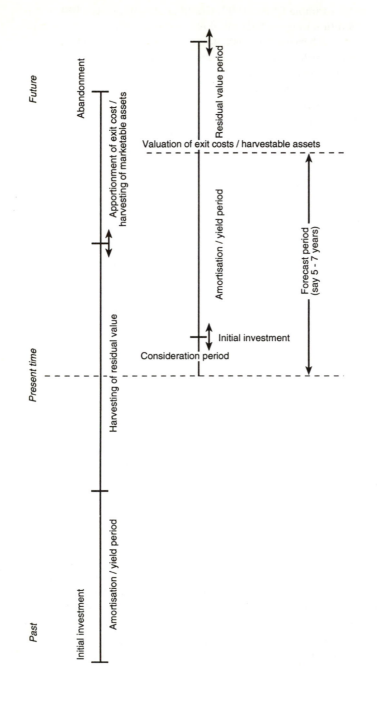

Figure 12.2 Why divisional managers find it difficult to get funds and why investment cycles are shortening

294

continued operation of the Newcastle steelworks remains viable while ever divisional revenues exceed recurrent costs. This is the key to the process of harvesting of residual value in a fully amortised plant.

The influence of market structure

A crisis in its accumulation strategy draws a corporation's attention to its market position as well as to the management of its production capabilities. As a result of the restructuring of BHP's Steel Division, the Newcastle steelworks provides feedstock for RBPD's Newcastle mills, and for BHP mills in Geelong (Vic), Kwinana (WA) and Brisbane (Qld). More than any other BHP Steel division, RBPD has been forced to adopt specific geographic strategies in its (mostly unsuccessful) attempts to secure market share, maintain steel throughput at the Newcastle works, and meet the corporation's earnings and profitability requirements. Because of the world-wide expansion in the number of mini mills, its product range is the most over-supplied internationally. When, in 1984, the Smorgon Group, a family-run manufacturing conglomerate, commenced steelmaking at an EAF mini mill at Laverton North in the industrial belt west of Melbourne (Vic), it was the first time BHP Steel had faced a genuine domestic competitor in raw steel production since its takeover of Australian Iron and Steel P/L at Port Kembla in 1932. Smorgon penetrated the Australian rod and bar market rapidly and pushed its market share towards thirty per cent as customers relished the opportunity to purchase steel at competitive prices.

The need to defend a significantly reduced market share following the Smorgon move drove RBPD's capital investment schedule during the decade that followed. Locationally, it reflected Hotelling's (1929) analysis of the defensive posturing of duopolists ever watchful of the other. In the late 1980s and early 1990s, the Smorgon Group made a number of strategic acquisitions of inter-state fabricators and building materials suppliers in an aggressive bid for national market share. RBPD countered with its own downstream takeovers drawing heavily on its meagre divisional investment allocations.

RBPD was also wary of potential new entrants. In the mid 1980s, Queensland's conservative Bjelke-Peterson government publicly sought expressions of interest in the establishment of a mini mill in the state capital city, Brisbane. BHP was (again) disinterested in the proposal since it would mean reducing throughput at its Newcastle operations. However, the establishment of a rival consortium, Brisbane Mini Steel Mills Ltd, to investigate a Brisbane mini mill spurred BHP investment. When the consortium announced plans to purchase a second-hand EAF from Toulon (Fr.), BHP acted quickly to protect market share and announced plans for the establishment of a market mill in Brisbane, to be fed by billets from Newcastle. The rival EAF proposal was dropped leaving RBPD holding even higher levels of unwanted capacity. The game continued in 1988 when BHP announced plans to construct its own EAF mini mill at Rooty Hill in Sydney's outer western suburbs in defence of its domestic rod and bar market position in Australia's largest city. BHP claimed that the establishment of a mini mill at Rooty Hill would allow more of Newcastle's production to be directed to the export of SBQ steel. Yet, output from the Rooty Hill mini mill displaced domestic supply from the Newcastle steelworks forcing RBPD to expand *opportunistic* exporting at marginal cost into an international rod and bar market already flooded with product from both mini mills and ageing integrated steelworks.

The production network which now constitutes BHP Steel's RBPD is largely the geographic manifestation of the division's monopolising market behaviours. Capital has been apportioned on the basis of securing domestic market share and erecting new barriers to entry, rather than on technical change or on the establishment of a genuinely *strategic* export strategy such as took place in other BHP Steel divisions. Yet, BHP's public statements concerning options for the future of the Newcastle steelworks have focused exclusively on technical matters. To have discussed the relative merits of different product ranges in different markets would have risked demolishing the carefully constructed public perception that the Newcastle steelworks' greatest problem related to international competitiveness. The propagation of this myth by BHP in conjunction with the public transmission of uncertainty concerning the future of the steelworks gave BHP considerable leverage in its demands for intensification of work at the plant:

> We've spent a lot of money at Newcastle in the past several years, which I would have thought should have been interpreted as a commitment to Newcastle as a steelmaking site. It is necessary, right at this time, for additional steps to be taken at Newcastle to restore the viability of the plant. The plant will downsize very substantially *if it loses its international market.* (BHP chief executive John Prescott quoted in *The Business Review Weekly,* 2.8.91, p.61, emphasis added)

BHP never refers to domestic competition in publicity surrounding the future of the Newcastle steelworks. The existence of Smorgon Steel is never conceded. The absence of secured overseas markets is denied.

The political relationship between the corporation and the state

Throughout the twentieth century, the expansion and restructuring of BHP steelmaking have relied on state guarantee of monopoly access to resources, capital subsidy, the provision of infrastructure, product market protection and the public funding of community dislocation. In examining the future prospects for the Newcastle steelworks, BHP continued to give close attention to its political strategy. A range of long-term options for the Newcastle steelworks was evaluated by RPBD in 1992 and reported in an internal document, *Future Directions Study.*

A key section of the study, the *Stakeholder Perceptions Analysis,* contains an investigation into the political impact of the considered options (BHP Steel RBPD, 1992).

The *Stakeholder Perceptions Analysis* involved two stages. The first stage was an evaluation of the reactions of various individuals, groups and institutions ('stakeholders') to eight hypothetical steelmaking or closure options contained in the *Future Directions Study.* The second stage involved devising politically-sensitive methods of communicating an options decision to the stakeholders. In the first stage, a matrix was constructed and stakeholders' responses to various technical and closure options were scored. Stakeholders were those with a significant interest in RBPD's restructuring decision. These included the then Australian prime minister, Paul Keating, the leader of the opposition parties, federal politicians, BHP shareholders, RBPD's Newcastle workforce, the local Newcastle community, and national and local media. The stakeholders' reactions to various BHP options were scored and analysed.

The process indicated BHP's belief that the Newcastle steelworks had a symbolic importance beyond its physical role as a steel producing plant. It has historically dominated Newcastle's economy, strongly affected the divisions of labour in a significant proportion of the paid and household workforce, sustained regional linkages, and veiled the aesthetic appearance of the city. More significantly, it symbolised BHP's commitment to Australian manufacturing in general, the Newcastle steelworks being Australia's first and longest surviving major manufacturing facility: the 'mother plant'. A withdrawal from Newcastle would signify both a departure from a community which supplied BHP with labour and profits for over eighty years, and a depreciation of the corporation's commitment to a renewed national manufacturing base.

By 1992, it became apparent to RBPD management that its headquarters would not fund a new steelmaking investment cycle at the scale of the existing Newcastle plant using new technologies. Yet, the *Stakeholder Perceptions Analysis* warned against the national political consequences of closure. The recommended option was to continue to harvest residual value. In implementing the recommendation, BHP Public Affairs has carefully managed media observation of the plant's future. When closure speculation grew in 1994, a successful public relations campaign managed to secure a local page one banner headline, 'BHP Steel in for the long haul' (*Newcastle Herald,* 28.5.94, p.1). Not surprisingly, concern over the future of the ageing works (including smoky coke ovens and poor health and safety record) prompted a further round of media massaging by BHP and resulted in a belated announcement of closure of the blast furnaces in 2002 and the installation of (unspecified) EAF technology to supply raw steel to the site's forming and rolling mills. Despite an announced 3,000 to 1,000 fall in employment during this period, BHP was able to sell the announcement as a commitment to its Newcastle operations. The *Newcastle Herald* obliged with the headline 'New era for steelworks' (1.7.95, p.1). In contrast, national newspapers reported the BHP announcement as simply

297

involving the continued operation of the old plant at lower staffing levels (*Sydney Morning Herald*, 6.6.95, p.2; *The Australian*, 1.7.95, p.1). In July 1996, however, commitment to the EAF proposal was qualified by the requirement that its viability be assured by yet another review of BHP's Australian steelmaking operations, and the 15-years-long process of pending closure continued.

The existence of the *Stakeholder Perceptions Analysis* illustrates BHP's inclusion of political factors in its decision making processes. But why does a large and powerful corporation have to be concerned about political repercussions from a decision to close a plant where the closure could be publicly supported on sound technical and economic reasons? One answer is that the corporation is mindful of the state's role in creating competitive advantage for the Petroleum and Minerals groups and for other BHP Steel divisions. Despite BHP's growing offshore investments, the company's output remains substantially based on Australian production spaces. BHP is reluctant to make decisions which upset these opportunities. The decision to close the Newcastle steelworks has been delayed until the political costs of closure are able to be absorbed or have diminished to a level where they can be ignored. Political costs will fall as employment numbers at the steelworks diminish. Further, they will fall as the steelworks adopts a lower profile in internal and external images of Newcastle. In this regard, RBPD Public Affairs Unit has adopted a new media strategy which involves intervention to *prevent* publicity about the steelworks so as to lower community concern about its future and diminish BHP's local profile to match the reduced size of the steelworks (RBPD Public Affairs management (*pers. comm.*, 11.11.92).

The public transmission of the 'internationalisation' discourse

BHP places great importance on its public information campaigning. In the 1980s, RBPD Public Affairs unit successfully steered Newcastle residents to believing that the steelworks' chief problem was a lack of international competitiveness:

> RBPD conducts an annual survey of the Newcastle community attitudes towards BHP. From this data basis, the annual public affairs strategy is laid. The external strategy is based around a 3 minute television segment. Surveys for 3 years have found that the majority of Newcastle residents think the most important issue BHP has to deal with in the 1990s is international competitiveness. The survey also reveals that there is little understanding of the harsh realities behind becoming internationally competitive. Therefore, RBPD has concentrated on increasing the public's knowledge of the challenges facing RBPD in the future. (*Future Directions Study*, September, 1992)

The control over public information is designed to use the image of internationalisation to gain leverage over organised labour to extract cost cutting concessions. The continuous presentation of uncertainty in general staffing levels, plant viability and international competitiveness, however accurate, has provided RBPD with an effective tool for an ongoing appropriation of productivity improvements throughout the 1980s and 1990s. The average productivity level for BHP's integrated steelworks rose from 180 tonnes of raw steel per employee year in 1983 to over 500 tonnes a decade later.

Six tools supported the intensification of the work process, RBPD's only avenue to improved profitability after adopting its strategy of harvesting the steelworks' residual value. The first tool was the threat of closure and job insecurity and coincided with the expiry of the Steel Industry Development Agreement (SIDA) in June 1992. Negotiations for a replacement for SIDA were conducted during the period of the *options* study when RBPD was refusing to eliminate closure as an option. Further, RBPD withheld job security guarantees until the final stages of the National Steel Industry Business Improvement Agreement (NSIBIA) negotiations, at times publicly threatening to withdraw guarantees if unions failed to comply with intensification requests.

The second tool arose when RBPD sought greater returns from its *residual value* strategy by engaging outside consultants (McKinsey), to implement a 20 per cent cost cutting programme using procedures previously used in intensification projects in steelworks in Germany and France (*Rodbar,* July 1991, p.5). The programme was called the Business Improvement Program (BIP) and it aimed to produce annualised savings of A$150 million by 1995, a large part of which derived from 1,400 job reductions from total RBPD employment but chiefly from the Newcastle works. These were achieved by the imposition of the Voluntary Early Retirement (VER) scheme whereby all employees aged over 50 years were offered early retirement, transfer to another department, or to 'special projects' such as cleaning and maintenance.

Business plans were the third tool. In the absence of strategies, besides harvesting residual value, RBPD's business plans concentrated on the adoption of performance targets in two areas, costs and productivity. The moves to enterprise and site agreements enabled the business targets to be formalised in agreements under the Steel Industry Plan, SIDA, and NSIBIA, as well as in BIP, thereby linking the achievement of management goals directly with workers' pay, job security and the long-term viability of the steelworks.

The fourth tool was the application of value added management (VAM) practices. Both cost cutting and productivity growth were enhanced following RBPD's adoption of VAM practices. Having established business units with defined boundaries for the calculation of costs, revenue and productivity, business units were then given the responsibility for achieving the goals determined for them by senior management. The process began in 1985 with the creation of separate steel, petroleum and minerals groups, and thereafter the separate product divisions in BHP Steel, and in 1989 the development of a VAM-based management structure within RBPD.

The use of the internationalisation image was the fifth tool. The claim that productivity had to be improved to match world best practice because of RBPD's alleged internationalisation strategies was consistently advanced by RBPD to both BHP workers and to the Newcastle community:

> The major challenge as I see it is to make the Division internationally competitive," he [RBPD general manger Paul Jeans] said. "I see this involving a continuation, escalation and intensification of the work currently being planned and undertaken by the RBPD management team. (*Rodbar,* February 1991, p.1)

The image of internationalisation served to justify massive job shedding by inculcating the harvesting of residual value strategy with a simultaneous appeal to nationalism and the creation of an ethereal, offshore enemy which was most to blame for job loss and intensification.

The sixth tool involved the use of spatial divisions of labour to speed the introduction of work place reforms. It included, first, the promotion of rivalry between the different unions covering the Newcastle and Port Kembla steelworks; second, competition between Port Kembla and Newcastle management to implement BIP; and third, the use of greenfields site agreements at the new mills in Brisbane and Rooty Hill as yardsticks for productivity performance at Newcastle and to force the introduction of contract work arrangements in areas such as maintenance, canteens, carpentry, painting, auto service, refrigeration and on-site transport. Under a strategy of harvesting residual value, contracting allows staffing levels to be eroded without labour shortages, and minimises employee severance payments in the event of plant closure. Not surprisingly, over ninety per cent of labour force cuts under RBPD's BIP programme were from the Newcastle steelworks site.

The influence of multiple class positions of local management

To understand the processes by which a steel division selects a new accumulation pathway, both the division and the corporation need defining in ways different to the organic and totalising metaphors most commonly employed in accounts of manufacturing change (O'Neill and Gibson-Graham, forthcoming). Resnick and Wolff (1987) divide the enterprise into a series of accumulation *and* distribution processes which together are seen to overdetermine intra-corporate relations:

> ...the enterprise, industrial or not, is the site of a set of distinct economic and non-economic processes. As the site of the economic process of surplus value appropriation, the capitalist industrial enterprise exists as a

particular location of the performance and appropriation of surplus value. As the site of the economic process of surplus value distribution, it also exists as the location of other economic processes such as purchasing of labour power, raw materials, and equipment and of non-economic processes such as supervising, commanding, advertising, and public relations. Individuals occupy specific class positions within the enterprise because they participate in either or both the fundamental and subsumed class processes specified to take place there. (p.167)

Yet individuals also participate in many other social processes which are not internal to the enterprise:

For [this] reason, individuals participating in these other social processes are said to occupy class and non-class positions external to the industrial enterprise. (Resnick and Wolff, 1987 p. 167)

A decentred conceptualisation of the enterprise and its managers is better able to explain the tensions between the management of a steel division motivated to maximise the success of a narrow steel product range, and corporate management which seeks to maximise returns to shareholders and thereby prefers investments in particular outputs over others. In addition, class positions held by steeldivision managers are overdetermined by processes from within the community surrounding a steelworks.

The tensions produced by managers' multiple class positions have played a critical role in investment decisions affecting the Newcastle steelworks. There has always been friction between local management and head office (Melbourne) management of the BHP corporation over the conduct of operations at the Newcastle works. During the period when the Newcastle steelworks was the engine room of BHP's steel production, its managers enjoyed a powerful stature, locally (in the works and in the community) and within the corporation. This stature was acknowledged by the corporation when it conducted rounds of promotion and transfer. The expected movement of the Newcastle general manager was to a senior management position in Melbourne. This was also in the corporation's interests since it removed the influence of class positions sourced within the local community from internal *modus operandi*.

Historically, there was a standard pattern: steelworks general managers elevated their class allegiance to the BHP corporation after promotion and placement in another city, Melbourne, away from the sites of steel production and within the ranks of those most trusted with the interests of capital. Table 12.1 describes the succession of general managers at the Newcastle steelworks. The two retired general managers were the only exceptions to the historical pattern of promotion. One, David Baker, the first general manager, was brought to Australia from the USA as an expert in the establishment of a steelworks and remained at the steelworks until his retirement. The other, John Risby, was the general

Table 12.1
Succession of general managers at Newcastle steelworks

Place of initial appointment to the BHP corporation		Service at Newcastle steelworks prior to appointment?		Destination following tenure as Newcastle steelworks	
Newcastle	9	Yes**	14	Promotion	13
Broken Hill	2	No*	3	Retirement	2
Port Kembla**	2			Termination**	1
South Australia	2			Currently GM*	1
Western Australia*	1				
Melbourne	1				

Notes:

*   denotes position of current general manager, Robert Kirkby
**   denotes position of general manager who commenced *options* study, Rob Chenery

Source: various BHP publications

manager at the steelworks during the early 1980s recession. He publicly expressed his community-based class interests during the consideration of closure in 1982 by campaigning for the continuance of steelmaking in Newcastle because of its social and economic importance to the local community and the regional economy. He allowed himself to be appropriated by the region's push for continued steel production within Newcastle and defied his internal capitalist class position which required the most profitable allocation of capital for maximum shareholder returns.

Risby's assistant general manager during this period was Rob Chenery. Having joined BHP at Port Kembla. Chenery was promoted to general manager of BHP Steel Collieries in the mid-1980s and, after a short appointment in Melbourne, became general manager of RBPD in 1989. He knew BHP's corporate culture intimately yet was ultimately denied the opportunity to cash in his multiple positions as most before had done by promotion to Melbourne. This denial was closely related to Chenery's role in internal contests over restructuring plans for the Newcastle steelworks. One of the tasks Chenery undertook as general manager of RBPD was the establishment of business units. In promulgating the changes, Chenery drew staff attention to two aspects of the reorganisation. First, he stressed that it was part of a long-term project of change, and secondly, it would '...position RBPD to attract significantly more capital investment...' (*internal RBPD memo*, 1.6.89). Chenery was aware that the steelworks required replacement or major refit and knew that RBPD had to meet BHP corporate criteria for the attraction of the necessary funds, estimated at $500 million. Chenery proceeded to focus RBPD's corporate plan so the division could achieve the required investment funds. The 1990-93 plan devised productivity and sales targets which matched the corporate goal of 15 per cent per annum on capital invested. Thereafter, Chenery began the *options* study confident that he would gain access to the investment funds needed for a substantial change to the steelworks productive capital. The demands of his multiple class positions could be jointly satisfied with a revitalised, secure Newcastle steelworks returning the required dividends to BHP shareholders. Chenery occupied his external class positions with relish. He was a key member of a number of community organisations, including a board member of the NSW government's Hunter Economic Development Council (HEDC), and was promoted publicly by organised labour as having genuine concern for the local community. In parallel, the steelworks unions publicised and supported Chenery's preferred option for the installation of a new technology, direct-iron-reduction, smelting plant at Newcastle, which would secure steelmaking in the city for another two decades (*Business Review Weekly*, 24.7.92).

Recession, the product constraints of the 1985 restructuring, and failing, old equipment ensured that the desired returns on existing investment were not achieved. Unfortunately for Chenery, rejection of his investment bid coincided with a round of promotions and transfers among senior BHP managers following the resignation of chief executive officer Brian Loton. Loton's replacement in January 1991 by BHP Steel chief executive John Prescott maintained the steel route of succession to the throne. At the next management level, however, the historical route was abandoned when Chenery was by-passed as head of BHP Steel. Prior to the public announcement of the new team, Chenery was advised that he had been passed by. He chose to resign immediately. He became the first Newcastle general manger to do so and took employment as head of the US Wheeling Pittsburgh Steel Corporation (*Business Review Weekly*, 2.10.92, p.24). He had been unable to reconcile the demands of the corporation with his multiple class positions. He had adopted a business plan whose measurable goals, if successful, would have achieved substantial capital investment for the Newcastle steelworks. His failure to attract promotion was a failure of coincidence between corporate and local community goals. He had elevated his local position above the class interests of the corporation and was denied entry to the level where the interests of capital are required to subsume all other class positions. In leaving Newcastle he said that '...the future of Newcastle Steelworks was in the hands of our (sic) own people' (*Newcastle Herald*, 30.1.91, p.2).

The lesson of Chenery's failure was not lost on his successor Paul Jeans. Jeans publicly acknowledged the poor performance of the Newcastle steelworks claiming its current rate of return was insufficient to justify substantial capital investment and abandoned the bid. Jeans elevated the class interests of BHP capital over local class interests, yet appeased the latter with a clever political and communications strategy. His Newcastle tenure was rewarded thereafter by appointment to general manager of the Port Kembla steelworks, the new core activity of BHP Steel, and, in 1995, to a senior management position in Melbourne. Jeans replacement at Newcastle, Robert Kirkby, was a further indication of BHP's concern to avoid its Newcastle manager having competing class positions. Kirkby had never been to Newcastle prior to his appointment, nor had he worked in steel. His self-professed expertise is in securing new work practices on behalf of capital, a skill developed in his previous position as general manager of BHP's iron ore operations at Mount Newman, a remote mining town in Western Australia.

The isolation of the RBPD general manager from some of the class positions traditionally occupied by BHP management in Newcastle culminated the abandonment of the regional linkages which BHP had maintained since 1915. This diminished regional role has been responded to politically with a conscious programme by RBPD Public Affairs Unit to reduce the importance placed on the steelworks by the community. Instead, publicity surrounding the steelworks has been consciously moved to the issue of steelworker productivity.

Conclusions

The detailed case study of restructuring at the Newcastle steelworks highlights the way forces from many scales intersect at the *site* of changes to work and capital investments. The analysis here is driven by the recognition that, at one level, we must dismantle the overwhelming discourse of the global. Yet, at another level, we must incorporate global scale process more accurately into our analysis. A major difficulty is that our understanding continues to grow *astride* the understanding of global management by key decision makers such as corporate treasurers and divisional managers - with their own revisions and reconstructions of the stories of change.

Certainly, there are new logics for undertaking investment. New explanations (however flawed) are available to corporate managers which link (rather than oppose) financial and productive investments. Technical evaluations, risks and yields from proposed new investments are assessed alongside ventures in other sectors and against the returns available from financial placements. The mechanics of this assessment seem to advance the strategy of the harvesting of residual value from existing large scale investments rather than their replacement by new investments (where the assessment process seems to preclude the calculation of residual value).

Opportunities for entry into new investment cycles, and the restructuring or continuation of old ones, are dependent also on market structures. It should not come as a surprise that the capital accumulation processes for monopoly capitalists have had something to do with the creation of specific market structures. Hence, reconsideration of the capital base will inevitably involve reconsideration of the market position. Again, stability in a market position seems to advance the desirability of intensification and a harvesting of residual value strategy. The same argument can be advanced in consideration of the political context of restructuring involving nationally significant monopoly capitalists. BHP places immense importance on stable political relationships with governments. With the rise of regional politics, large corporations are forced to consider the political outcomes of exit decisions. The case study demonstrates that this political evaluation process can be formalised and documented - to exist as a competing story of change alongside technical and financial options.

The stories of change constructed within the corporation, then, drive the restructuring process. For the corporate treasurer, the story is one involving the forecast period, perhaps a brief five to seven year period which is causing major contractions in investment cycles, diverting financial capital away from long term productive investments and producing the rationale for the continuance of the harvesting of residual value from older plant. For the divisional manager the stories produce a dilemma. This person has the task of providing information to the corporate treasurer as the basis for a request for finance for a new investment cycle. Yet, the divisional manager must also tell a story of restructuring to secure (or coerce) the cooperation of a local workforce and community; and the divisional manager knows the historical processes that have led to the complex inter-relationships between physical capital and the workers who operate it. Certainly, the story of *internationalisation* can serve as a powerful message to extract productivity increases. But it can also lead to contradiction in management actions with managers often unsure as to where to place their divided allegiances.

Restructuring is a complex international *and* local event. It takes a form determined by the creation of logics by corporate decision makers. Removing the blinkers of totalising 'globalisms' helps the researcher discover these logics. That they are bound by the history of existing investment cycles, and by technical, financial, market and political constraints provides the opportunity for the construction of competing stories and logics with more favourable social outcomes.

References

Archibald, D. (1991), *The Australian Steel Industry,* Smith New Court, Sydney.

Australian Steel Industry Authority (1988), *Review of the Steel Industry Plan. A Report to the Minister for Industry, Technology and Commerce,* Parliament House Canberra, June.

Baran, P.A. and Sweezy, P.M. (1966), *Monopoly Capital: An Essay on the American Economic and Social Order,* Penguin, Hammondsworth.

Baumol, W.J., Panzar, J.C. and Willig, R.D. (1988), *Contestable Markets and the Theory of Industry Structure,* 2nd edition, Harcourt Brace Jovanovich, San Diego.

BHP Ltd, various years, *Annual Report,* BHP, Melbourne.

BHP Ltd, various years, *Factsheet,* BHP, Melbourne.

BHP Ltd, various years, *Pocketbook,* BHP, Melbourne.

BHP Steel (1986), 'The Steel Industry Plan - Half Way', *News Release,* BHP Steel, Melbourne, 15 July.

BHP Steel (1989), The Broad Allocation of Factors Influencing the Steel Industry's Position, Internal document, BHP Steel, Melbourne.

BHP Steel, RBPD (1992), *Future Directions Study,* internal document, Newcastle.

Bradbury, J.H. (1987), 'Technical change and the restructuring of the North American steel industry', in Chapman, K. and Humphrys, G. (eds), *Technical Change and Industrial Policy*, Oxford, Basil Blackwell, pp. 157-173.

Chew, D. (1993), *The New Corporate Finance,* McGraw Hill, New York.

Clark, G.L. (1988), 'Corporate restructuring in the steel industry: adjustment strategies and local labor relations', in Hughes, J. and Sternlieb, G. (eds), *America's New Economic Geography*, Centre for Urban Policy Research, Rutgers University, New Brunswick.

Clark, G.L. (1992), 'Strategy and structure: corporate restructuring and the scope and characteristics of sunk costs', paper presented to a meeting of the Institute of Australian Geographers Study Group on Industrial Change, University of Newcastle, NSW, September 30 - October 1.

Clark, G.L. and Wrigley, N. (1993), 'Sunk costs: a framework for economic geography', paper presented to the annual meeting of the Association of American Geographers, Atlanta, Georgia, 7 - 9 April.

Clark, G.L., McKay, J., Missen, G. and Webber, M. (1992), 'Objections to economic restructuring and the strategies of coercion: an analytical evaluation of policies and practices in Australia and the United States', *Economic Geography*, vol. 68, pp. 43-59.

Dicken, P. (1992), *Global Shift: Industrial Change in A Turbulent World,* 2nd edition, Harper and Row, London.

Dicken, P. (1994), 'Global-local tensions: firms and states in the global space-economy', *Economic Geography*, vol. 70, pp. 101-128.

Dicken, P., Forsgren, M. and Malmberg, A. (1994), 'The local embeddedness of transnational corporations', in Amin, A. and Thrift, N. (eds),*Globalisation, Localisation*: *Possibilities for Local Economic Prosperity*, Oxford University Press, Oxford, pp. 23-45.

Dicken, P., Peck, J. and Tickell, A. (1997), 'Unpacking the global', in Lee, R. and Wills, J. (eds), *Society, Place, Economy, States of the Art in Economic Geography*, Edward Arnold, London, in press.

Donaldson, M. (1981), 'Steel into the eighties: the rise and rise of BHP', *Journal of Australian Political Economy*, vol. 10, pp. 37-45.

Donaldson, M. and Donaldson, T. (1983), 'The crisis in the steel industry', *Journal of Australian Political Economy*, vol. 14, pp. 33-43.

Fagan, R.H. (1984), 'Corporate strategy and regional uneven development in Australia: The case of BHP Ltd', in Taylor, M. (ed.), *The Geography of Australian Corporate Power*, Croom Helm, Sydney, pp. 91-124.

Fagan, R.H. (1986), 'Australia's BHP Ltd - an emerging transnational resources corporation', *Raw Materials Report*, vol. 4, pp. 46-55.

Fagan, R.H. (1988), 'Corporate structure and Australian industry: a geography of industrial restructuring', in Heathcote, R.L. (ed.), *The Australian Experience*, George Allen and Unwin, Sydney, pp. 25-36.

Fagan, R.H. and O'Neill, P.M. (forthcoming), *'Globalism', locality and the state: the internationalisation of BHP*, unpublished paper.

Fisher, J.R. and Smith, A. (1980), 'International competition in the Australian wire market 1880-1914', *Business History*, vol. 22, pp. 71-86.

Garlick, S. and Skaines, I. (1983), 'Regional employment and economic impacts of the reduction in BHP steelmaking operations in Newcastle', *Research Report, No. 40*, Hunter Valley Research Foundation, Tighes Hill, Newcastle.

Goold, M., Campbell, A. and Alexander, M. (1994), *Corporate Level Strategy: Creating Value in the Multibusiness Company*, John Wiley, New York.

Harvey, D. (1982), *The Limits to Capital*, Blackwell, Oxford.

Hotelling, H. (1929), 'Stability in competition', *Economic Journal*, vol. 39, pp. 41-57.

Hudson, R. (1989), 'Labour-market changes and new forms of work in old industrial regions: maybe flexibility for some but not flexible accumulation', *Environment and Planning D: Society and Space*, vol. 7, pp. 5-30.

Hudson, R. and Sadler, D. (1986), 'Contesting works closures in Western Europe's old industrial regions: defending works or betraying class', in Scott, A.J. and Storper, M. (eds), *Production, Work, Territory: The Geographical Anatomy of Industrial Capitalism*, Unwin Hyman, Boston, pp.172-194.

Hughes, H. (1964), *The Australian Iron and Steel Industry 1848-1962*, Melbourne University Press, Melbourne.

Larcombe, G. (1983), 'Corporate strategies or economic planning? The future of Australia's industrial regions', in Gordon, M.T. and Gordon, B.L.T. (eds), *Policy Priorities for Australian Steel Regions*, Institute of Industrial Economics, University of Newcastle, NSW, pp.47-63.

Lewis, J.E. and McDonald, R.R. (1986), 'The Australian steel industry: new directions', paper presented to CHEMECA 86, Adelaide, 19-22 August.

Love, R.W. (1995), 'Shareholder value analysis: the linkage between finance and strategy', *The Australian Corporate Treasurer*, February, pp. 9-13.

MacLachlan, I. (1992), 'Plant closure and market dynamics: competitive strategy and rationalisation', *Economic Geography*, vol. 68, pp. 128-145.

Mandel, E. (1978), *Late Capitalism*, translated by J. De Bres, New Left Books, London.

O'Neill, P.M. (forthcoming), 'Internationalisation and the nation state: Australia's changing accumulation strategy', *Environment and Planning A*.

O'Neill, P.M. and Gibson-Graham, J.K. (forthcoming), 'Stories that destabilise the company: class analysis and the enterprise', in *Class: The Last Post-Modern Frontier*, Routledge, New York.

Resnick, S.A. and Wolff, R.D. (1987), *Knowledge and Class: a Marxian Critique of Political Economy*, University of Chicago Press, Chicago.

Rich, D.C. (1986), *Industrial Geography of Australia*, Methuen, Sydney.

Scherrer, C. (1991), 'Seeking a way out of Fordism: the US steel and auto industries', *Capital and Class*, vol. 44, pp. 93-120.

Schultz, J. (1985), *Steel City Blues: The Human Cost of Industrial Crisis*, Penguin, Melbourne.

Schumpeter, J.A. (1947), 'The creative response in economic history', *Journal of Economic History*, vol. 7, pp. 149-159.

Stafford, H.A. and Watts, H.D. (1991), 'Local environments and plant closures by multi-locational firms: a cross-cultural analysis', *Regional Studies*, vol.25, pp. 427-438.

Webber, M., Clark, G.L., McKay, J. and Missen, G. (1991), *Industrial restructuring: definition*, Working Paper 91-3, Monash-Melbourne, Joint Project on Comparative Australian-Asian Development, Monash University Development Studies Centre, Melbourne.

Index

310

107, 116, 117,120, 126-128, 130, 132, 134, 135, 137-141, 147, 148, 150-157, 159, 161, 163-166, 169-174, 176, 178, 182, 196, 197, 204, 217-219, 221, 267-271, 278

institutions, 4-5, 10, 18, 22-26, 31, 33,35, 37, 39, 44, 46, 49, 58-59, 66-68, 72-73, 75-76, 82, 87, 91, 115-116, 122, 127-129, 139, 142, 147, 150-151, 153, 155, 159-161, 163, 165-166, 170-173, 175, 177-178, 183, 185, 196, 199-200, 212-213, 215-217, 233, 258-259, 267, 276, 280, 297

integration, 13, 20, 35, 36, 38, 39, 47, 57, 59, 61, 66, 70, 71, 77, 86, 91, 93, 104, 111, 112, 144, 145, 147-149, 156, 158-160, 164, 166, 182, 185, 188, 189, 209, 215, 227, 257, 258

internationalisation, 9, 13, 16, 19, 20, 26, 28, 29, 65, 66, 86, 102, 109, 111, 113, 115, 144, 179, 190, 229, 230, 235, 283-287, 290, 292, 299, 300, 305, 307

investment, 17, 45, 76, 86, 89, 101-103, 107, 111-113, 130, 134, 173, 184, 185, 199, 200, 213, 230, 231, 236, 242, 245, 248, 253, 254, 256, 266, 270-272, 284, 289, 292, 293, 295, 297, 301, 303-305

isomorphism, 72

Japan, 17, 43, 51, 88, 91, 102, 134, 159, 168, 170, 175, 196, 221

knowledge, 3, 6, 8, 18, 24, 27,

35-39, 41, 46, 47, 51, 53, 58, 61, 64, 70, 75, 78, 83, 90, 91, 106, 107, 111, 113, 117, 118, 120, 123, 126-135, 139, 148, 149, 151, 153, 161, 162, 164, 166, 168, 178, 187, 197, 198, 200, 201, 213, 218, 219, 249, 298, 308

Korea, 249, 275

Kwinana, 288, 295

labour, 3, 2, 4, 9, 20, 26, 33, 35, 38, 46, 49, 54, 61, 62, 65, 67-70, 72, 73, 75, 77, 90, 94, 99, 102, 112, 122, 134, 136, 142, 144-146, 150, 155, 159, 160, 167, 170, 171, 175, 176, 179, 183, 184, 186, 189, 193, 196, 202, 216, 220, 221, 223, 224, 226, 227, 229-236, 241-245, 247-251, 253-281, 283-290, 297, 299-301, 303

labour markets, 9, 61, 67-70, 73, 75, 77, 112, 122, 150, 186, 223-224, 227, 229, 232-233, 236, 241-245, 247-251, 254-255, 257-265, 267-274, 276

liberalisation, 3, 286

localisation, 84, 85, 98, 121-123,136, 142, 148, 176, 254, 257, 259, 261, 262, 267, 271-274, 307

management, 12, 19, 25, 32, 36, 46, 52, 53, 61, 62, 64-67, 70, 73, 76, 78-80, 82, 83, 88-91, 94, 96, 97, 102, 130, 139, 140, 150, 161, 162, 171, 174, 179, 184, 186, 189, 193, 196, 219, 230, 254, 256, 257, 259, 260, 263, 265, 267, 269, 284-288, 293, 295, 297-301, 303-305

manufacturing, 31, 52, 60, 89, 105, 127, 133, 138, 145, 158,